Relatively and Philosophically Eᵃrnest

Festschrift in Honor
of Paul Ernest's 65th Birthday

THE MONTANA MATHEMATICS ENTHUSIAST
MONOGRAPH SERIES IN MATHEMATICS EDUCATION

Series Editor
Bharath Sriraman, The University of Montana

Relatively and Philosophically Eᵃrnest

Festschrift in Honor of Paul Ernest's 65th Birthday

edited by

Bharath Sriraman
The University of Montana

Simon Goodchild
University of Agder, Norway

INFORMATION AGE PUBLISHING, INC.
Charlotte, NC • www.infoagepub.com

Library of Congress Cataloging-in-Publication Data

Relatively and philosophically earnest festschrift in honor of Paul Ernest's 65th birthday / edited by Bharath Sriraman, Simon Goodchild.
  p. cm. – (The Montana mathematics enthusiast: monograph series in mathematics education)
 Includes bibliographical references.
 ISBN 978-1-60752-240-9 (pbk.) – ISBN 978-1-60752-241-6 (hardcover)
1. Mathematics–Study and teaching. 2. Mathematics–Philosophy. 3. Constructivism (Philosophy) 4. Ernest, Paul. I. Sriraman, Bharath. II. Goodchild, Simon, 1950-
 QA11.2.R44 2009
 510.1–dc22

                                                    2009030518

Cover photo: Paul Ernest on the Oval at The University of Montana, Missoula with Main Hall and Mt. Sentinel in the background. ©2007 Bharath Sriraman

Printed in the United States of America

CONTENTS

CHAPTER 1

SOCIALLY (RE) CONSTRUCTING PAUL ERNEST

Bharath Sriraman
The University of Montana

It is an honor and privilege for me to able to compile and edit this book together with Simon Goodchild, to celebrate the 65th birthday of Paul Ernest. My familiarity with Paul Ernest's work began in graduate school in the mid '90s when I started to read the works of Lakatos. One of the early reviews of *Proofs and Refutations* was written by Paul Ernest (see Ernest, 1978). His name started to appear more frequently when I began looking at the "boundaries" of proof in mathematics. Upon trying out mathematics education in a graduate survey course, his name became even more ubiquitous for one interested in learning the foundations of the field. Interestingly enough I was rebuked by one newly hired assistant professor for proposing that his book *The Philosophy of Mathematics Education* was much more relevant than the other disconnected and boring readings we were being made to digest in this foundations course.

I had seen Paul at some meetings, but we met in the real sense of the word at a European mathematics education meeting in Spain about five

Relatively and Philosophically E[u]*rnest*, pages 1–8

Figure 1.1   Paul Ernest in Montana in 2007 (Mission mountains in the background).

years ago. "Boredom" again played a role, as Paul and I were simultaneously trying to escape the orthodoxy being propagated by some colleagues in some working groups. This led to several long walks and the discovery of a kindred spirit and mutual interests that had led us down similar paths (geographical and metaphysical) in our youth albeit two decades apart! The similarity in our pasts led us easily into a friendship, and I am naturally honoured to be counted by Paul among his close friends. In 2007, Paul spent his Easter vacation with us in Montana, at which time the idea of having a Festschrift for his 65th birthday as part of the Montana Monograph Series dawned in my mind. I am pleased that with the help of the other authors that have contributed to this Festschrift, the book has indeed materialized!

In this opening chapter I will socially (re)construct Paul Ernest (pun intended) for the reader unfamiliar with his life, his contributions to the philosophy of mathematics and the field of mathematics education.

THE FORMATIVE YEARS OF PAUL ERNEST

Paul Ernest was born in New York September 23, 1944, son of a Jewish-American artist John Ernest and Swedish psychologist Elna Adlerbert. Paul's father was always known as Ernie, although his name was Jachiel Cohen, which he changed for a number of reasons including the avoidance of anti-Semitism. John Ernest was later to become a founder member of the British constructivist art movement in the 1950s and1960s. His work is highly prized and is represented in museums and private collections throughout the UK, Europe and, the USA, including a number of works in the Tate Gallery, London. John was also a gifted amateur mathematician and discovered a mathematical result in graph theory. Elna Ernest worked as a clinical psychologist until her retirement in 1980 and now paints full time and lives in Spain.

In 1946 Paul and his parents moved first to Gothenburg, Sweden and then to France in 1949, where they lived in Paris and Antibes. The family moved to London to settle permanently at the time of the Festival of Britain in 1951. By this time, 6-year-old Paul was trilingual. The family lived in West Hampstead and Paul attended both primary and secondary schools in the Hampstead area, including William Ellis Grammar School in Parliament Hill. The 1960s was an exciting time for growing up around Hampstead, and Paul aspired to the Beat and protest Generation, participating in four annual anti-nuclear Aldermaston marches as well as many peace, anti-Vietnam war and anti-apartheid demonstrations during the early 1960s. Much of his youth was misspent at the infamous Belsize Park coffee house, the Witches Cauldron. He later became a fellow traveller of the Mod, and later the Hippie movements in the mid 1960s. He attended live concerts by Tamla Motown artistes, the Beatles, Rolling Stones, Miles Davis, John Coltrane, Thelonious Monk, Jimmie Hendrix (5×), Pink Floyd (4×), Captain Beefheart and others now forgotten.

Although, ostensibly a student of science and mathematics, Paul spent much of the 1960s reading modern and classical literature and poetry, philosophy, mythology and mysticism, in the appreciation of modern painting, classic film and early music (medieval to baroque), and learning meditation and yoga, as well immersion in the popular culture of the time.

PAUL—THE "E(A)RNEST" SEEKER

Like many of his generation, Paul Ernest dropped out from his studies in the 1960s, and then travelled extensively in Europe, Morocco, Turkey, the Levant, Iran, and Afghanistan. In between bouts of study he held a number of jobs including refuse collector, security guard and computer programmer— designing and installing the first computerized accounting system for H.M. Treasury, London in 1971. He met Jill, his wife to be, in 1970, and they were married in 1972. Jill undoubtedly contributed to his new found seriousness and resumption of study. In 1971 he graduated in mathematics, philosophy and logic from Sussex University in 1973; was awarded an MSc with distinction in mathematical logic by London University in 1974; and finished his doctorate on the philosophy of mathematics and logic in 1984.

Paul was accepted to study for an undergraduate degree entitled Philosophy and the theory of science (with mathematics) at Sussex University, in the Logic Division of the School of Mathematical and Physical Sciences. By the time he graduated it was called Logic with Mathematics. The degree fundamentally remained the same with 50% pure mathematics and 50% philosophy of mathematics, philosophy of science, the history and philosophy of logic, epistemology, and some other scientific aspects of modern philosophy. The teachers with the greatest impact on Paul were Jerzy Giedymin (philosopher of science) and Yoshindo Suzuki (mathematical logic).

For his masters degree Paul was registered at Bedford College, University of London, situated in Regent's Park. His tutor was the well known mathematical logician and historian of mathematical logic, G. T. Kneebone. Although, he took his main courses at Chelsea College (mathematical logic, proof theory, recursion theory with Moshé Machover) and London School of Economics (Boolean algebra, model theory, axiomatic set theory with John Bell).

For his doctoral studies, Paul was a student in the Department of the History and Philosophy of Science at Chelsea College of Science, London University, with supervisor Moshé Machover. By the time the PhD was awarded, Chelsea College had become part of King's College London, and the HPS Department was absorbed into the Philosophy Department.

For his doctoral studies, Paul made an in-depth study of mathematical logic, the foundations of mathematics and the philosophy of mathematics and developments in these fields from the 1870s to the 1970s. His doctoral thesis was entitled 'Meaning and Intention in Mathematics', and was based on using proof theory (semantic tableaux) and model theory to provide a formal definition of the meaning of mathematical expressions with both intensional (proof based) and extensional (truth based) components. The thesis defined a measure of the distance between the meanings of any two mathematical expressions, and proved the equivalence of a number of formal relations involving meaning, such as synonymy, and meaning distance 0. It also drew on Wittgenstein's meaning as use doctrine.

Paul began his studies in the 1970s with a traditional Platonist/absolutist philosophy of mathematics. His studies of Intuitionism, Lakatos and Wittgenstein led him to begin to question this philosophy. However, it was the immersion in mathematics education, first as a mathematics teacher 1976–1979, and then more decisively as a mathematics teacher educator 1979 and on, that led to the full humanisation of his philosophy of mathematics ultimately leading to his development of a social constructivism. It also led to his highly critical stance towards absolutism in philosophy of mathematics, arguing forcibly against his own beliefs of a decade earlier that he had subsequently repudiated. A first version of his social constructivism drawing on radical constructivism was published in 1991, and the fully developed social constructivism as a philosophy of mathematics using Vygotsky's work and semiotics appeared in print in 1998.

Since 1976 he has devoted himself full time to the teaching and learning of mathematics and to mathematics education, beginning with a 3 year stint as a mathematics teacher in a London comprehensive. This was followed by a series of positions in mathematics teacher education at college/university level. Becoming a mathematics teacher educator started changing Paul's interests towards to mathematics education, as he reflected on pedagogical issues and mathematics curriculum in order to teach these subjects to trainee teachers. It also accelerated the humanization of his philosophical views. Then after 4/5 years as a teacher educator reflection on problem solving, investigational work and other pedagogies he realized how important he believed people's personal philosophies were for their practice, and that his earlier studies in philosophy of mathematics were a valuable resource to apply in reflecting on and theorizing teacher belief systems. Paul attended his first international conference in 1986—PME10 in London. For the next ten years he attended every annual PME conference religiously. This served as a Summer school, introducing Paul both to the leading edge theorizing in mathematics education, and also to many of the leading authors and researchers in the field personally.

FROM SEEKER TO GATEKEEPER

After a few papers on relations between the psychology of learning mathematics and mathematics education, teachers attitudes and beliefs, and the mathematics curriculum making connections, he wrote his widely cited 1991 book (*The Philosophy of Mathematics Education*) in 12 months, from around Easter 1989 to Easter 1990. He has never–before or since–been so driven!

Paul worked as a mathematics teacher educator at Homerton College, Cambridge 1979–81, and Bedford College of higher education, Bedford, 1981–82. He enjoyed an exciting two years as lecturer in education at the University of the West Indies, Jamaica, 1982–84, where he designed a distance education course for mathematics teachers which, until recently, was still transmitted around a dozen Caribbean territories by satellite. He took up his present post at the University of Exeter first as a lecturer in 1984. He was promoted to reader in 1994, and full professor in 1998. Paul Ernest is currently Emeritus Professor of the Philosophy of Mathematics Education. He is also a visiting professor of mathematics education at Oslo University (first appointment 2006–2009), and HiST-ALT College, Trondheim (first appointment 2007–2010).

He has taught and developed student materials at all levels, from undergraduate teacher education, Postgraduate Certificate of Education (primary and secondary), through to Masters Degree and doctoral degrees in Mathematics Education (both EdD and PhD). He also chaired the development of the overall research methodology programme attended by all doctoral students in education at Exeter University. He developed and directed the specialist doctoral and masters programmes in mathematics education taught in a unique distance learning format that attracted students from almost every continent. For these programmes he authored doctoral course handbooks on:

1. The psychology of mathematics education
2. The Mathematics Curriculum
3. Mathematics and Gender: The Nature of Mathematics and Equal Opportunities
4. Mathematics and Special Educational Needs
5. Research Methodology in Mathematics Education

Paul Ernest's research addresses fundamental questions about the nature of mathematics and how it relates to teaching, learning and society. His academic interests include:

- The relationship between the philosophy of mathematics and mathematics education, and more generally the philosophy of mathematics education, ethics and values in mathematics education, and the philosophy of research methodology
- The mathematics curriculum and its aims, curriculum ideologies, etc
- Critical mathematics education and all aspects social justice in mathematics education including issues of gender, race, special educational needs, class, etc
- Psychology of mathematics education and learning theory in mathematics education
- The semiotics of mathematics education and language in mathematics education
- His earlier publications also explored mathematics teaching pedagogies and the use of information and communication technologies and computers in mathematics education.

His best known publications include the books: *The Philosophy of Mathematics Education*, Routledge, Falmer 1991, and *Social Constructivism as a Philosophy of Mathematics*, SUNY Press, 1998, plus the edited books *Mathematics Teaching: The State of the Art* (1989), *Mathematics, Education and Philosophy:*

An International Perspective (1994), and *Constructing Mathematical Knowledge: Epistemology and Mathematics Education* (1994).

Paul is the founder editor of the international book series, *Studies in Mathematics Education,* with Routledge/Falmer press. In 1990 Paul founded the *Philosophy of Mathematics Education Journal (PoME),* freely available at http://people.exeter.ac.uk/PErnest/. It has included special issues on semiotics and mathematics education, social justice in mathematics education and mathematics and art. Paul's conception of a journal being freely accessible to the community served as an inspiration for me to found The Montana Mathematics Enthusiast (TMME) in a similar spirit, and keep it free for the community. It is thus apt that Paul's Festschrift is part of the monograph series that arose out the success of the journal. We were also recently able to collaborate on editing a book with Brian Greer on Critical Issues of Mathematics Education released in July this year, which synergized the efforts of PoME and TMME, into a monograph of substance and interest to the community. Visit http://www.infoagepub.com/index. php?id=9&p=p490a29296b4d1.

To date Paul has published over 200 books, chapters and papers in learned journals, and given over one hundred presentations at national and international conferences in countries including: France, Spain, Denmark, Italy, Hungary, Sweden, Norway, Finland, Belgium, Greece, Cyprus, Abu Dhabi, USA, Mexico, Canada, Brazil, Japan, India, Malaysia, Brunei, Thailand, Barbados, Bermuda, England, Scotland and Wales.

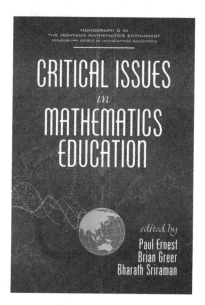

Figure 1.2 Monograph 6 of TMME

Figure 1.3 Ernest and Sriraman, Ravalli, Montana.

Paul enjoys theorizing and, some would say risky speculation, about philosophical aspects of mathematics and mathematics education and their role in society. The wide range of topics covered in the chapters of this Festschrift bear testimony to his wide array of interests. Last but not least, Paul Ernest also has numerous personal interests including computing/internet, academic writing, cooking/food and wine, gym/keeping fit, literature, poetry and popular fiction, cinema, art/museums, music from popular, blues & world to classical, travel/sightseeing, scuba diving (especially diving with sharks). It is with great pleasure I join the other authors in this Festschrift in thanking Paul for his contributions to the mathematics and mathematics education community, his intellectual honesty and spirit, and in wishing him a happy 65th birthday.

REFERENCE

Ernest, P. (1978). Review of 'Proofs and Refutations' by I. Lakatos, *Mathematical Reviews*.

CHAPTER 2

LISTEN TO YOUR SUPERVISOR!

Simon Goodchild
University of Agder, Norway

INTRODUCTION

This essay is written in celebration of an effective doctoral supervisor. The purpose is to reflect on some experiences arising from my good fortune to have Paul Ernest as my supervisor, and in doing so, place on public record my gratitude for his contribution to my work. Universities devise forms of agreements, contracts, and codes of practice for the relationship between supervisor and PhD student. Typically, these will include details about frequency of supervision, responsibilities, and deadlines and the like. For reasons that, I guess, are rather self evident, the codes of practice that I have seen do not include statements about the supervisor's knowledge or the competence of a supervisor to put that knowledge to use effectively in supervision or guidance. Nor do those codes of practice define the lasting bond of friendship that can emerge between a supervisor and student, or a measure of debt the successful PhD holder feels towards his/her supervisor. Those fortunate to have enjoyed a fruitful and enriching relationship with a doctoral supervisor will recognise the importance of these qualities and outcomes, an importance that transcends the instrumentality of "codes

*Relatively and Philosophically E*ⁿ*rnest*, pages 9–17
Copyright © 2009 by Information Age Publishing

of practice." As I reflect here on the impact of Paul's supervision on my doctoral research I believe I am drawing attention to these essential but uncodifiable qualities of supervision.

THE SUPERVISOR'S GOAL IS TO ACHIEVE INVISIBILITY!

Since completing my PhD at the University of Exeter in 1997, I have had the opportunity to introduce Paul at a number of different events. Each time it has been my pleasure to recognize Paul's creative touch on my work, and each time with growing insistence Paul has modestly denied my acclaim. However, I recall the anecdote of a man who passes a wonderful garden that is delightfully landscaped with well-clipped lawn, shrubs and flowers, and not a weed in sight. In admiration the man remarks to a woman, who is busy tending to one of the beds, "Isn't nature wonderful in producing such beauty." To which the woman replies: "Well, you should have seen its chaotic state when left to nature!" Fortunately, very few people saw the chaos of the first draft of my dissertation. In this essay I will recall some episodes from my experiences as Paul's PhD student, which I hope will demonstrate his ability to create some order from chaos.

BACKGROUND

I worked for my PhD in my spare time, alongside a full time job at what was then The College of St Mark and St John (Marjon),[1] Plymouth. It is about 40 miles between Plymouth and Exeter. Meetings with Paul were fairly infrequent, once or twice each semester, they were always scheduled and generally rather long. At the time there was an arrangement between Marjon and the University of Exeter through which Marjon academics could register without cost on the university doctoral programme. This was a good deal for me and my colleagues at Marjon; I am not so sure it was so good for the supervisors at Exeter because this 'pro bono' activity was not recognized as part of their regular work load. As with many such institutional arrangements the price is ultimately paid by hard working and dedicated staff, in my case it was Paul who *gave* time for the supervision of my work. When I started on the PhD I had in mind, given that it was a spare time project, I would probably take 10 to 12 years to complete. Paul would not agree to this, he wanted to see it completed in much shorter time and in fact it was less than five years between the first meeting I ever had with Paul and the *viva voce*. I believe that Paul was absolutely right to insist that the work did not drag on for too long. It ensured the work remained fresh and alive.

However, I do not think it would have been achieved without Paul's ability to focus my attention on what I was doing.

FOCUSING THE INQUIRY

Two particular episodes stand out in which I especially recall Paul's focusing influence, once at the start and the other near the end of my doctoral work. I first outline how he managed to direct me into a fruitful line of research, this took place in the very first meeting I had with him in the late spring of 1992. I had submitted a proposal for PhD research to the University of Exeter, and fortunately it was passed to Paul, who invited me to meet him and talk about it. The proposal was naïve and far from Paul's interests, it was quickly put aside and Paul embarked on a "lecture" that I guess was intended to enable me to construct a better informed, and rather more sensible proposal. Also, I guess, one which suited Paul's own interests but here one must acknowledge that Paul's interests are very wide so this is hardly 'restrictive'. The personal lecture lasted about three hours and I left Paul reeling! As I drove back to Plymouth afterwards I felt as if I had been in the sights of a scholarly machine gun firing ideas and references— with no means of cover or escape. I guess it is possible to be critical of the "pedagogy"; but, I recall it as an invigorating and exciting experience that made me want to continue. Amongst all the ideas, names and references that Paul "fired" at me that afternoon were Stieg Mellin-Olsen and the book "Perspectives in Mathematics Education" (Christiansen, Howson & Otte, 1986) both were new to me. They became the principal theoretical informants of the classroom research in which I embarked. As I reflect now on *my* meetings with new or prospective students I think, and hope, that I may be a bit gentler than I experienced Paul that afternoon; however, I doubt the experience for the students is so exhilarating or motivating. As I write this I also speculate about which of the books I recommend to beginning students may have a similarly focusing effect on their work.

Throughout the time in which I enjoyed Paul's supervision I was continually impressed by his apparently encyclopaedic knowledge of the literature of mathematics education and a wider associated range of sources, particularly in sociology and philosophy and cognitive psychology. The early meetings with Paul were generally unidirectional; I would have sent something to Paul for him to read, which would be quickly critiqued and set aside—nevertheless it created an agenda for a seminar in which I hardly dared to argue. However, that did not remain the situation for very long. As I became better informed through the literature Paul was recommending me to read, and read more widely as I followed cross-references, it became possible to begin to engage with Paul, perhaps not balanced, but certainly in dialogue. That was until the day that I expressed some ideas about which Paul admitted he did not have superior knowledge. And he listened, and questioned, and challenged! I know it is the case with the students whom I supervise that there comes a time when their knowledge outstrips my own—it generally comes rather sooner in the process than in my association with Paul! At this point the supervisor needs to exercise some humility. I am grateful to Paul that he exercised the required humility; he listened, and took the ideas I expressed seriously and helped me to work on them and develop them.

FOCUSING THE DISSERTATION

During the spring of 1996, as I was drafting chapters for the dissertation, I became increasingly aware of impending changes in the working conditions at Marjon that would make sustained effort in the PhD more difficult. I should add, that throughout my time employed at Marjon I received great support especially from my closest colleagues in the Mathematics Subject Group (Richard Harvey and Roger Fentem), and from senior managers. The support was real in terms of the generosity of easing my workload, and covering costs to attend international conferences, secretarial support and research assistants. The support was also less tangible in terms of the encouragement and enthusiasm of colleagues. The completion of the dissertation, in my spare time, within five years is due in large measure to Paul's support and focusing; but, it was also thanks to colleagues at Marjon. However, Marjon was not operating in a vacuum and the external pressures being exerted on teacher education programmes had consequences on the support that could be given to staff pursuing higher degrees and research activity.

At the end of May 1996 I sent a draft of the first three chapters (Introduction, theory and methodology) to Paul. It was clear in my mind that with these chapters in place the remainder would be achievable by mid-September, when the new semester would begin. As noted above, whenever I met with Paul my practice was to send him something to read in advance, leave it two or three weeks and then call him (pre e-mail days) to arrange a meeting. My belief was then, and remains, that Paul did not read the papers I had written until after I had arranged the meeting. Perhaps I am being unfair but I "learned" that a doctoral student must know how to manage his or her supervisor and I learned to "manage" Paul! I arranged to meet with Paul to discuss the chapters I had sent. They were the first draft chapters that Paul had seen. When we met in late June it was to find out that he did not like them! His message was clear, revise Chapter 2 and make it Chapter 1, revise Chapter 1 and call it Chapter 3, write a completely new

Chapter 2—based on a line of literature that was completely new to me. I believe he had not managed to get as far as to read the draft methodology chapter.

I guess most that have experienced doctoral supervision will be able to recognize the inner sinking feeling I had as I pondered on the enormity of the additional work that was being required. Obviously my dismay was evident to Paul because later that evening he called me at home to encourage me to continue—it was the only time he ever took the initiative to call me. The call, perhaps because it was so unusual, was very important to me then and it now tells something about Paul's sensitivity as a supervisor, to know when to leave alone and when to intervene. As above, I want to give credit to others as well as Paul for the encouragement I received at that time, and in particular I want to acknowledge Nora Linden's support. Paul had introduced me to the written work of Stieg Mellin-Olsen. It is a great sadness to me that I never met Stieg Mellin-Olsen who died in January 1995. However, Marilyn Nickson arranged for me to meet Nora, Stieg's widow, so that I could discuss my interpretation of Stieg's work. I had sent Nora a copy of the chapters that Paul condemned, and after the meeting with Paul I wrote to tell her about Paul's reaction. Almost as soon as she received my letter Nora called me from Norway to encourage me in doing what Paul wanted.

Fortunately for me one of the new references that Paul wanted me to read was Noel Entwistle's book, in which he writes about deep and surface approaches to learning (Entwistle, 1981). I began to read the book "instrumentally" to enable me to write the chapter Paul required, but as I read I could see that I should apply Entwistle's message to my own situation, to use Mellin-Olsen's term, with an "S-rationale" (Mellin-Olsen, 1987). This text transformed my own attitude towards what Paul was asking me to do, and helped to bring into being the chapters that he was asking for. Within six weeks of that meeting with Paul the three new chapters were in place, and I was back to the position in which I thought I was nearly three months earlier.

Paul's advice, however, was far more crucial than just delivering the first chapters from my chaotic weed strewn thinking in the first half of the dissertation. His advice provided a structuring resource (Lave, 1988) for the rest of the dissertation. I have, on other occasions, reflected on the significance of Paul's focusing, and remarked rather casually that the rest of the dissertation *wrote itself*. What I mean is that the structure and focus that Paul had perceived in the new chapters (one, two and three), focused my mind on the data that I had spent over 18 months analyzing, enabling the chapters discussing the analysis and interpretation to flow. Several years after completing the thesis I could still read the interpretation chapters and get a sense of excitement from them, and for this I thank Paul for his perception and ability to focus the work I was doing.

I regret I cannot recall who was speaking to give the credit for an observation I heard recently; the remark related to supervision that was so heavy and demanding that at the end of the day the "student" took no pleasure in her/his dissertation because she/he did not recognize it as her/his own work. The comment made me reflect on my experience with Paul, especially the episode I have described above and wonder whose work my dissertation really is. It has never occurred to me before that it might be anyone else's but my own. Paul's supervision was focusing and enabling, it was critical and demanding, but at no stage could I describe it as heavy, oppressive, or insistent to write *his* account. Now that I find myself a supervisor of doctoral students, I reflect often on how I benefited from Paul's guidance and I aspire to enable and encourage "my" students, as he enabled and encouraged me, without crushing their independence or agency.

VALUING OTHER VOICES

I do not think Paul ever felt there was any threat to his reputation if I should choose to listen to, and perhaps follow the lead given by, other influential researchers. Indeed, Paul encouraged me to engage in the national and international communities, and it was Paul who introduced me to many of the international leaders in our field. I believe Paul would be the first to admit that his strength lies in theoretical interpretation and development, rather than fieldwork and data collection. His strengths were complemented by another person, to whom I feel a great debt in the development of my PhD work and as a researcher; this is Leone Burton whom I met at meetings of the British Society for Research into Learning Mathematics. I learned a lot from Leone about the meaning of methodology but I must still note important lessons I learned—or adapted from Paul. I believe one of Paul's intentions was that I would identify myself as a researcher. On several occasions, as we discussed issues relating to the field work I was intending or doing, Paul would remind me that I was a researcher in this context—not a teacher! This was quite a hard lesson to learn as I had been teaching at various levels for over twenty years by that time, and it was through my experiences of teaching mathematics that I was motivated to research teaching and learning mathematics in classrooms.

I agree with Paul that there is a difference between entering a classroom as the teacher and as a researcher—no matter what the agreement is between the teacher and researcher (Golby & Appleby, 1995; Wagner, 1997). Nevertheless, and I do not know if Paul would argue the point. I do believe that it is significant, and I would claim an advantage, for a mathematics classroom researcher to have an extensive background as a mathematics teacher. Paul's point was that the two roles should not be confused and that the researcher's first priority was to observe, to pursue the research goals, and to collect authentic data. The researcher is not the teacher and it is both poor research practice and possibly overstepping the boundaries of agreement between researcher and teacher, to slip out of the researcher role and into the teacher's.

In my PhD research the agreement I had with the teacher and students was that of "data extraction" (Wagner, 1997). I attended a class on a regular basis to collect data. I talked with students because the conversations were the main source of data. I talked with the teacher as a matter of courtesy. But my role was not to teach the students, nor was it to collaborate with the teacher in the sense of holding up a mirror to her practice. Ten years on I question the ethic of the position I was supposed to occupy. I write "supposed" to occupy because despite Paul's exhortation to the contrary there were times when I believe it was appropriate to "slip" into the role of a teacher, and I did. If a student made a direct request to me for help, it would have been inappropriate for me to make a flat refusal, and then a little later ask the student to talk with me on my agenda. The response I took to such requests was to say that I am allowed to ask questions, not answer them, but that I was prepared to ask some questions that might prove helpful. I guess most readers will recognize this as teaching. Also, in conversation with students, and there were many over the course of the year I spent with the class, there were times I became aware of their errors or weaknesses in their understanding. Here I want to assert that it is inappropriate to leave a student without drawing attention to an error, which may be interpreted by the student as implicit approval of her or his flawed product. In such cases, during the conversation I would try to frame questions that might lead the student to recognize that something was incorrect—indeed this was part of the process of exploring the student's understanding in the spirit of teaching experiment methodology (Steffe & Thompson, 2000). If this failed, I would advise the student to seek help from her/his teacher. I believe, in these responses, I was overstepping the boundary that Paul wanted me to observe. The teacher did not on any occasion request that I comment on any aspect of her lesson, our agreement did not include this and the teacher appeared to respect my researcher role.

I have moved even further from the position that I believe Paul defined for me as researcher. For the last five years I have been engaged in mathematics teaching developmental research. The agreement that I seek with teachers Wagner (1997) characterizes as "co-learning" and as I work with teachers and students I occupy a dual role in both developmental and research activities. I believe that educational research is developmental research; one purpose for the research is that it makes a difference. On the other side, without research change cannot be counted as development. I have written more extensively on this theme (Goodchild, 2008), the point is that in the type of classroom research that I have learned to value, the extensive experience I have as a teacher, both to inform the research and equip me personally for the role is vital. I believe that one of the important lessons I have learned from Paul is to keep an open mind—and that includes being open to disagree!

SUMMARY (DEFINITELY NOT A CONCLUSION)

It is my wish that everyone who has completed a PhD could look back on a very special relationship with the supervisor afforded by the experience. I guess that for many, the few experiences I have described above from my time as Paul's student could be matched in some way. Paul, I am sure would modestly insist that he was not "special." However, I had only one principal supervisor and from my perspective it was a special privilege to work with a scholar of Paul's quality and reputation. I am convinced that his ability to focus me in my research and to challenge my thinking made a crucial contribution to my development as a researcher. I now have the opportunity to meet with many doctoral students through the courses I teach at the University of Agder in Norway, inevitably there are occasions in which I express opinions that are contrary to those of the students' supervisors. Out of respect for their supervisor and consideration for the student, I lay down one rule that I am sure comes from my experience with Paul: "Listen to your supervisor!"

Perhaps I should add, "But, you do not necessarily have to agree with everything he (or she) says."

NOTE

1. Present name: University College Plymouth St Mark & St John.

REFERENCES

Christiansen, B., Howson, A. G., & Otte, M. (Eds.) (1986). *Perspectives on mathematics education*. Dordrecht, Holland: Reidel.

Entwistle, N. (1981). *Styles of learning and teaching*. Chichester, UK: John Wiley & Sons.

Golby, M., & Appleby, R. (1995). Reflective practice through critical friendship: some possibilities. *Cambridge Journal of Education, 25*(2), 149–160.

Goodchild, S. (2008). A quest for 'good' research. In B. Jaworski & T. Wood (Eds.), *International handbook on mathematics teacher education: Vol. 4. The mathematics teacher educator as a developing professional: Individuals, teams, communities and networks* (pp. 201–220). Rotterdam, Holland: Sense.

Lave, J. (1988). *Cognition in practice: Mind, mathematics and culture in everyday life.* Cambridge, UK: Cambridge University Press.

Mellin-Olsen, S. (1987). *The politics of mathematics education.* Dordrecht, Holland: Reidel.

Steffe, L. P., & Thompson, P. W. (2000). Teaching experiment methodology: Underlying principles and essential elements. In A. E. Kelly, & R. A. Lesh (Eds.), *Handbook of research design in mathematics and science education* (pp. 267–306). Mahwah, NJ: Lawrence Erlbaum.

Wagner, J. (1997). The unavoidable intervention of educational research: A framework for reconsidering research-practitioner cooperation. *Educational Researcher 26*(7), 13–22.

CHAPTER 3

NEW WINDS BLOWING IN APPLIED MATHEMATICS

Philip J. Davis
Brown University

How would you complete this sentence: "This is The Age of..."?

Every generation of writers has filled in the dots not only for their own age, but for selected past ages. In 1947, the poet W. H. Auden wrote that his was "The Age of Anxiety." Around 1970, song writers, pulling on astrological beliefs, were cheerful (and whistling in the dark) called it "The Age of Aquarius," an age of love and human kindness. More recently, psychiatrist Daniel Freeman wrote that this is the "Age of Paranoia" with distrust and optical surveillance everywhere. We answer according to our experiences.

I would fill in the dots by saying that this is the computer age, or more sharply, the age of mathematizations. The computer is the prime and driving mechanism of the age and behind all computer applications there resides some sort of mathematical construction, chipified or otherwise. It is an age when applied mathematizations affect us all, for good, for bad, for somewhere in between, and these effects may not develop or become apparent for some while.

Mathematics is now so universally employed that its teaching cannot be encompassed in one department. CAD/CAM (computer aided design and

Relatively and Philosophically E^{a}rnest, pages 19–22

manufacture) is now in dentistry. Did the dentistry development require engineering talent? Should its techniques be taught in an engineering department? One can truly wonder what courses should comprise the basic training for the applied mathematician/computer scientist.

I have spent a good fraction of my professional life in what might be termed, a traditional department of applied mathematics. By "traditional" I mean a department that stresses the mathematics that models physical phenomena, or to a lesser extent that models social phenomena via statistics. The word "traditional" can also be explicated by noting the specific courses that are given in such a department. In my department at Brown University, for example, there are graduate courses in biophysical models, genomics, operations research, statistical inference, dynamical systems, and fluids. This represents a change from a half century ago, when my department was a renowned research center for solid mechanics: elasticity, plasticity, rheology, etc.

The word "traditional" can also be explicated by the well known paradigmatic sequence:

Description → prediction → comparison → re-tinkering the description.

But there is now another type of applied mathematics whose paradigm is

Prescription → adoption → surveillance and societal evaluation → re-prescription

Here are a few simple examples of prescriptive mathematics that extend from single numbers to exceedingly complex systems:

The speed limit on a highway.
The mandatory retirement age for particular occupations.
The scoring system for football.
An algorithm for determining the "pecking order" of colleges.
The old "point system" for determining the quality of a mortgage application.
The presidential electoral system in the USA.
A national tax system.
National and international financial systems.
This list could go on and on.

Society may adopt a mathematical prescription, but its acceptance is more provisional or tentative than, e.g., Newtonian mechanics.

There is yet another kind of mathematics by prescription that derives from human behavior and that is the product of a large cadre of professionals whom the journalist and investigative reporter Stephen Baker has termed *The Numerati*. Who are The Numerati? A secret society? No, they are mathematicians, computer scientists engineers, physicists, economists, biologists, psychologists, linguists, data miners and sweeper- uppers of the personal habits of individuals. In fact, anyone who consciously devises or uses algorithms that extract patterns from the behavior of individuals, individual or collective, whose uses then have a direct impact on their personal lives. Admittedly this is rather vague, but it will clarify a bit as I go on and mention a few of the many examples that Baker gives:

- A name indentifying company using linguistic analysis can tell you whether Mr. Chang is the same fellow as Mr. Tchang, or even Mr. Tchung .
- A company keeps shopping and life style data of some 200 million Americans. "The company buys just about every bit of data about us that is sold, and then sells selections of it."
- "Another company quietly amasses court rulings, tax and real estate transactions, birth and death notices" so as to enhance, among other things, law and child support enforcement, public safety and health care.
- Yet another company divides the electorate into ten groups with characteristic voting patterns so as to help political parties get the swing voters onto their bandwagon.
- One of the Numerati "sees sensors eventually recording and building statistical models of almost every aspect of our behavior. They'll track our pathways in the house.... They'll diagram our thrashing in bed and chart our nightly trips to the bathroom."

Some of these compilations and analyses, even now, go on silently and automatically. We are typecast as we sit innocently before our screen and ogle, google and surf or when we use a credit card to buy our weekly groceries. The products of The Numerati can run from socially benign and useful to worrisome to scary. The principal worry is the loss of personal privacy.

What formal mathematical knowledge is required to install such mathematizations? It can be anything from the very simple to very advanced. Technically speaking, what goes on here is part of the field of learning/ computational statistics, a very hot area in applied mathematics, and computer science (CS.) The useful formal training would include: linear algebra, multi-variable calculus, optimization, probability, statistics, on the mathematics side. On the CS side, it would include algorithms and database theory. Specialized knowledge of the particular domain is of course necessary. In a sense, all applications of math are ultimately prescriptive, but I'm concerned here with mathematics that, once prescribed, *creates a brand new milieu* as opposed to mathematics that describes or models an existing often physical milieu.

In a perceptive 1946 article, the French polymath Paul Valéry, pointed out that with Volta's 1800 discovery of the electric current and invention of the battery, science entered a new phase wherein it created and described absolutely new phenomena as opposed to phenomena that pre-existed. By the same token, applied mathematics which has hitherto been concerned with pre-existing phenomena, is now, via the work of the Numerati, creating new (and largely social) phenomena. This adds fuel to the social constructivist view of mathematics as explicated by Paul Ernest in his many works.

Over the years, I may very well have trained students in applied mathematics who have happily, productively and lucratively, entered the ranks of The Numerati. There are fortunes to be made in the Numerocratic domains and the young people are aware of it. The low-hanging fruit in engineering may now have all been plucked. To go forward with schemes for, e.g., creating sufficient quantities of clean energy may be difficult whereas creating successful new data-mining applications is comparatively easy. This bodes well for the future of the numeratological field. As society moves from democracy to numerocracy, there is always much much more that can be done. And we must deal with its consequences, many of which will be unintended.

Mathematics is a very adaptable, very universally applicable language and for this reason it should be invoked with caution. Part of the mathematical education of the future should be to inculcate caution lest we fall into the complaint of Caliban: "You have given me language, and my profit on't is I know how to curse."

REFERENCES

Baker, S. (2008). *The numerati.* Houghton Mifflin.

Davis, P. J. (Sept., 1980). Mathematics by fiat ? *The Two-Year College Mathematics Journal, 11*(4), 255–263.

Davis, P. J. (2006). *Mathematics and common sense.* A.K. Peters.

Davis, P. J. (2009). New winds in applied mathematics: The road to numerocracy. *SIAM NEWS*, to appear: 2009.

Ernest, P. (1998). *Social constructivism as a philosophy of mathematics.*

New York: SUNY Albany Press.

Valéry, P. (1970). A *personal view of science,* in: *Occasions.* NJ: Princeton University Press.

CHAPTER 4

TENSIONS BETWEEN MATHEMATICS, THE SCIENCES, AND PHILOSOPHY

Jean Paul Van Bendegem
Vrije Universiteit Brussel
Center for Logic and Philosophy of Science
Universiteit Gent

INTRODUCTION

It is a safe guess to assume that the title of this article immediately suggests to the informed reader that he or she expects to be presented with a presentation of and/or discussion about the so-called *Science Wars*. This introduction will in fact briefly say a few things about this curious phenomenon, but it will not be the core theme of my presentation. Rather deliberately, I chose the term "tensions"—indeed, I did have Thomas Kuhn's *essential tension* at the back of my mind, see Kuhn (1977)—and not "wars," precisely to avoid this obvious connection. Apart from the fact that presently (August, 2008) the Wars seem to be in a phase of a (temporary?) armistice, it is my belief that the whole discussion has not been particularly productive, let

Relatively and Philosophically E^arnest, pages 23–37

alone seminal and that therefore it would be a good thing to leave it quite simply behind us.

These in a nutshell are the elementary facts. As is well-known the Science Wars started with the publication in 1996 of a paper, entitled "Transgressing the Boundaries: Towards a Transformative Hermeneutics of Quantum Gravity" by Alan Sokal (1996a) in the journal *Social Text* (although some will claim that the Gross-Levitt 1994 book set the scene for coming events). Shortly after publication, Sokal (1996b) announced in *Lingua Franca* that the paper was a hoax,[1] thus marking the true beginnings of war activities. Afterwards he joined forces—the military language does seem quite appropriate in this context—with Jean Bricmont, a Belgian physicist, which resulted in the publication of *Impostures Intellectuelles*. Roughly speaking, although with clearly distinct motives, in this book two scientists explain to us, laypersons, all the mistakes, stupidities, and sheer madness philosophers dare to write down and, in addition, consider worth publishing. Truth be said, some philosophers would indeed make a wise choice if they would cease to select their examples from the natural sciences and admit that mathematics is not exactly that branch of human knowledge they are familiar with.[2] The philosophers that are dealt with in the book are, though not exclusively, so-called post-modernists or, at least, so labelled.[3]

In short, once the Wars started, relationships between (exact) scientists and (some group of) philosophers were seriously disturbed. Scientists accused philosophers of not understanding what science is all about, of poisoning the youth—an accusation philosophers are actually rather proud of![4]—and, generally, of wasting people's and scientists' time and money, or, in other words, of having little or no economical utility or relevance—something actually philosophers are very proud of as well.[5]

However, there was and there still is something truly odd about this warfare business. Some of the possible participants were lacking, namely, the philosophers of science, including the philosophers of mathematics.[6] Although some attention is paid to them in the Sokal & Bricmont (1997) book, mainly towards Paul Feyerabend, the worst enemy imaginable in their view, they did not have the opportunity to really join into fights and battles. Whatever the reasons may have been for their reluctance, it did rather drastically change the whole debate. I am convinced that their contribution would have been vital, as it would have led not merely to an armistice, but also to a mutual understanding, enrichment, and support. To substantiate this claim, the space available is not sufficient, hence a different option has been taken. What follows is an imaginary dialogue between a philosopher, Paul,[7] and a mathematician, Martha[8] (occasionally interrupted by intermezzos to provide some background information). Why a mathematician and not a scientist *tout court*? Quite simply, because mathematics still enjoys this very special status of a "science" of certainty and reliability, and thus allows me to be a bit more explicit about these issues.

PHILOSOPHY AND MATHEMATICS: AN IMAGINARY DIALOGUE

We meet with Paul, a philosopher, and Martha, a mathematician, having the following discussion about the relations between their two domains of research:

P: Isn't it odd that scientists and I guess mathematicians as well have so much fun with stupid philosophers who don't understand the first thing about numbers and measures? After all, if it wasn't for us philosophers, would there be any mathematics at all? Without Plato and Aristotle, to name just these two, who would have told us what kind of special form of knowledge mathematics is? A source of certainty, of undoubtable truths, supported by an underlying, implicit logic, guaranteed to maintain the highest standard of certainty? So much for the ignored role of philosophers!

M: Steady on! Are you really trying to tell me that the Egyptians when they were measuring up the land using a 12-knot rope, did not know what they were doing? OK, so perhaps they could not prove that the equation $a^2 + b^2 = c^2$ has an infinite number of integer solutions, but they did know that $3^2 + 4^2 = 5^2$ comes out right. The mere fact that they accepted this technique for constructing a right angle, is "proof" in itself that they realised that doubting this technique would not do. Look: isn't it the same story over and over again? We, mathematicians, and, by extension, scientists have all the ideas and once they have been formulated, you philosophers come along and start to criticise us about the tiniest detail, thereby hampering our progress, something you fellows and girlies do not seem particularly worried about.

INTERMEZZO

Given the equation $a^2 + b^2 = c^2$, the integer solutions are given by the formulas: $a = p^2 - q^2$, $b = 2pq$ and $c = p^2 + q^2$. One easily checks that $(p^2 - q^2)^2 + (2pq)^2 = (p^2 + q^2)^2$. Take $p = 2$ and $q = 1$, and $a = 3$, $b = 4$, $c = 5$ is the well-known solution. What is not really trivial, is the other way round, viz., any integer solution can be expressed by these formulas. Note, in addition, that some questions can be answered in a rather straightforward way, once one has these equations. It follows, e.g., immediately that there are an infinite number of solutions. Although one has to be careful! In principle one needs to prove that for a different choice of p and q, the corresponding triples a, b, and c are indeed different as well. Take, e.g., the formula $n = ((2p + q) - (p + 2q))/(p - q)$, then it is not the case that, for different choices p and q, a different value for n will result. For, after simplification, one sees that $n = 1$, for all values of p and q.

P: Oh dear, surely you are not suggesting that one of the basic distinctions between science, mathematics in particular, and philosophy is that the former knows progress and the latter does not? Yes, I do know the basic argument: in mathematics problems get a generally accepted and no longer doubted solution, but is there one philosophical problem that is considered solved by all philosophers?

M: Although I was not going to use that argument, now that you have done so yourself, well, is there?

P: Tricky business this! If I were to answer this question, I would accept the presupposition of that question, namely, that progress in no matter what, means that you solve problems once and for all. This is simply far too narrow a view. Apart from the fact that even in mathematics problems are not necessarily solved for eternity, ...

M: Pardon? Oh God, no, you're not one of those post-modernists who think they can doubt anything?

P: Wait, wait, let me first finish my sentence: so, apart from that fact, there are many more interesting interpretations to understand progress. Just one example: given a problem, whether philosophical or scientific need not matter, suppose you are able to show that a proposed solution will not do. This implies that future generations need not waste time over that particular solution any more, so they can focus on other matters. That too I should want to count as progress. Wouldn't you agree?

M: Like I said, it wasn't my idea to use that argument, so, yes, why not, if you want to have progress, have it by all means! But, forget this silly matter, what did you just say about proofs not being eternal? That was a joke, right?

P: By no means! Haven't you read *Proofs and Refutations* by Imre Lakatos? Probably never even heard of the book, am I right? (Martha nods, somewhat annoyed.) Thought so! Well, all I can say, you really must have a look. The whole book centers around one particular "proof"—I have to use quotation marks, you'll see in a minute—and how it was changed and transformed over the years into something that bares almost no resemblance to the original attempt.

M: And what is the "theorem"? I guess I better use quotation marks myself.

P: Euler's "theorem"—you're quite right, by the way, about the quotation marks—says that, given a polyhedron, then $V(\text{ertices}) - E(\text{dges}) + F(\text{aces}) = 2$. Euler produced an ingenious "proof," there I go again, that can actually be summarized in words. Imagine a polyhedron. Each face can be triangulated by adding the necessary number of edges from one vertex to another. Now if you introduce such an edge, the number of edges goes up 1, but so does the number of faces and these two balance one another. So, it doesn't change the formula. Imagine now the triangulated polyhedron. Remove one triangle. What happens? We now have one face less and the same number of vertices and edges, therefore for this (amputated) polyhedron the formula now becomes $V - E + F = 1$. Now remove the triangles one by one in such a way that either one edge is free or two edges are free, in the sense that they do not belong to another face. In both cases, the formula remains unchanged. If one edge is free, you remove one edge and one face, which is OK. If two edges are free, you remove one vertex, two edges and one face and that too is OK. What will be left at the end of this procedure is one triangle and $3 - 3 + 1 = 1$. QED. Isn't this lovely?

M: To be honest, I would like to see a more formalized version of this nice story, if you don't mind. Are you quite sure about these free edges? I have a funny feeling something is missing there.

P: Excellent! Exactly what happened historically: imagine two polyhedra glued together at one vertex. For each one apart, you have $V - E + F = 2$, but for the pair of them, you end up with 4, minus the common vertex, hence $V - E + F = 3$.

M: But that's ridiculous, "glued together at a vertex"? That's not a polyhedron, that's plain silly! Get your definitions right, for Christ's sake! (Martha gets a bit angry).

P: My goodness, calm down! You know, what you have done just now is actually rehearsing the history of mathematics all by yourself. How strange!

M: Look, what I do know, is that, if you reformulate the problem in terms of vector spaces, graphs, linear transformations and incidence matrices, then a formal proof exists.[9] So what else have you shown besides the trivial fact that mathematicians need some time to get their ideas straight. But once straight, always straight. So your example is not all that impressive, is it?

P: Although even that is not entirely true. But, agreed, let us forget about this particular example. Just take a look at what we see happening today. I'll give you three examples: extremely long proofs, extremely complicated proofs and proofs involving computers, either for proof checking or for brute force calculations. Take as an example of the first, Wiles' proof of Fermat's Last Theorem; of the second, the proof of the classification of the finite simple groups; and of the third, the proof of the four colour theorem.

M: You have doubts about these proofs?

P: This might sound funny, but it doesn't matter for that is not the point I am trying to make. What I want to say is that these examples have radically changed the existing notion of proof , whereby I mean that something is a proof if, first, it is a finite sequence of formulas, second, any single mathematician can check this proof, and, three, it is part of a network of already proven formulas.

M: I see why the first two are important, but what is this idea of a network of formulas?

P: Well quite simply, the idea that, when a mathematician proves a theorem, he or she will make use of statements already proved without obviously repeating their proofs. This is so well-entrenched in mathematical practice that often mathematicians do not notice that they are using other theorems. It even goes further: often in the introduction of mathematical textbooks, there will be a paragraph summing up what will be assumed to be known.

INTERMEZZO

One of the finest examples I know of is to be found in Devlin (1988, p. 126). There he writes the following: "His idea was to concentrate on the elements a of the group (other than the identity element e) for which a*a = e. Such group elements are called involutions, and it is easy to show that any group with an even number of elements must contain at least one involution. (Try this yourself. All you need to know about groups is the definition given earlier. . . .)" The author of this paper of course immediately yielded to this invitation. The proof I came up with looks like this: Suppose that for every $a \neq e$, $a*a \neq e$. From this follows that for such $a \neq e$, $a \neq a^{-1}$. This follows from the fact that, if $a*a \neq e$, then $a^{-1}*(a*a) \neq a^{-1}*e$ or, by associativity and identity, $(a^{-1}*a)*a \neq a^{-1}$ or $e*a \neq a^{-1}$ and thus $a \neq a^{-1}$. This implies that the elements of the group can be split up in pairs (a, a^{-1}), so we must have an even number of elements. But for e itself, since $e*e = e$, so $e = e^{-1}$. So, we end up with an odd number of elements. Contradiction. Note the curious move here. I assumed that, if we have pairs (a, a^{-1}), then going through all the elements of the group, we must end up with an even number of elements. This is however (once more) trickier than it looks. Suppose two such pairs (a, a^{-1}) and (b, b^{-1}). If $a = b$, then $a^{-1} = b^{-1}$ and so the pairs coincide. No problem. But if $a \neq b$, then it must be the case that $a^{-1} \neq b^{-1}$, a fact that can be proven. Most importantly however, it also supposes that the sum of a set of pairs produces an even number. No doubt this will be considered a trivial remark, but it is a fact of number theory and not of group theory, hence the phrase "All you need to know about groups is the definition given earlier" is, strictly speaking, not correct. Actually, one needs a larger background than a mathematical background, but that is another discussion.

M: OK, although this all sounds rather obvious to me!

P: Perfect! For in that case you will agree that the three examples I just mentioned violate one or more elements of my definition, would you not? So, after all, it does not seem all that obvious after all.

M: What do you mean "violate"? Show me!

P: You want proof, I'll give you proof, to quote Sidney Harris.[10] Wiles first. When he published the proof, there were about ten mathematicians worldwide who could check the proof for its correctness. Today, because of important efforts of the mathematical community to simplify the proof, this number has considerably grown, but still it remains the case that not every mathematician can actually read the proof in the sense that he or she can read a proof in the first chapters of any handbook on whatever mathematical topic, where definitions are given and simple basic theorems are proved. But, as soon as things get "interesting," it is out of the question to present a proof in full detail as it would involve millions of lines, and the "real" proof requires so much mathematical background that only a part of the mathematical community can understand it and satisfy themselves as to its correctness. What are the other mathematicians supposed to do?

M: You call that a violation!? What is the problem? If a hundred mathematicians can understand and check the proof, that's fine with me. Your definition need not be a description of actual practice, does it? Sure, it would be much better if every mathematician could do the job, but a theorem does not become "more true" if, instead of one hundred, one thousand mathematicians can understand it (Martha laughs, as she find the idea of "more true" apparently very funny).

P: Really? Suppose that we have a proof and there is a mistake in it and suppose for simplicity's sake, that a mathematician on average has a probability p of detecting the error. In realistic cases, p will not be one. So there is a non-zero chance that he or she will not find the error, namely, $1 - p$. Suppose now that there two mathematicians who independently check the proof, then the probability becomes $(1 - p)^2$ and that is closer to 0. Take $p = 0.9$, then $1 - p = 0.1$, and so $(1 - p)^2 = 0.1^2 = 0.01$. Don't laugh, it does make a difference, alright?

M: Look, don't bother me with probabilities. A statement is true or not. If you have something that claims to be a proof, either it is a proof or there is a mistake in it, and then it is not a proof. There is no gradation of any sort, it is quite simply one or the other. Alright?

P: (Paul gets a bit irritated) Please do not confuse a proof being correct with somebody telling you that that proof is correct. That is the problem I am trying to explain: the only thing you can do is accept their judgment and, since they are human, they could be wrong. Unless you can check it yourself, which you can't. See?

M: Don't get so excited! Like I said, it is an ideal and, yes, sometimes we are closer to it, sometimes not. But it seems to work rather well, does it not?

P: Sorry, I apologize, but the point I am trying to make is really important to me, and I am really glad you mentioned the idea of "being closer or not." You will agree that, if we were able to measure this distance, that would be a good thing?

M: It certainly wouldn't cause any harm, so yes.

P: But to be able to measure this distance, you must know where the two points are that are separated by that distance and one of the two we know, namely, the ideal. Does that not imply that we therefore need to know the other point as well, mathematics as it is actually practiced?

M: My God, a real philosopher! Alright, yes! So do your measurements, if you like, what is it to me?

P: I am sure we will come back to this theme, but, in first instance, it is nothing to you. If you are doing mathematics, that's what you are doing. You read or are told that someone has given a proof of FLT, you say "Great," you hear colleagues who understand the proof that is a beautiful piece of work, you say "Great" a second time, and then you get back to your own specialty and continue to look for proofs for the open questions in your field. But an entirely different matter arises when you're asked to say something *about* mathematics, and you keep repeating the ideal and do not say a word about actual practice. Is it then not a nice "division of labour" that we philosophers take care of that job?

M: Sounds almost too good to be true. So you will not bother with us and just let us do our job. Fine with me.

P: Of course, the idea of the "division of labour" is itself more of an ideal than anything else (Paul clearly enjoys making this remark). Of course, philosophers will not sit in the corner of your office to study what you are doing while you are searching for proofs, but we will write and publish about it and then it is perfectly possible that you hear things that will shock you.

M: Try me!

P: My pleasure! Here goes: suppose I wrote an article, entitled "Correct proofs no mathematician can understand" in a philosophical journal. A colleague of yours shows you my paper. You see the title, what would your response be?

M: The truth? Obviously, that you don't understand the first thing about mathematics, something I suspected right from the beginning, to be honest. That is absolute nonsense: a correct proof can be understood. What's next: blue is not a colour?

P: Don't get upset! (Paul smiles). Suppose now that you read the paper nevertheless and you are confronted with the following argumentation. It will also allow me to say a few words about the second example I mentioned, namely, the classification theorem of finite, simple groups. When the proof of that theorem was considered complete, the complete version counted some fifteen thousand pages. Well, it is reasonable to claim that no *single* mathematician can check this proof. Can I tease you with a bit of logic? The statement "A proof that no mathematician can understand" is the statement "There is a proof such that it is not the case that there is a mathematician who understands it,"[11] and that is the only thing I am claiming. Agree?

INTERMEZZO

One of the fascinating aspects of the classification theorem is that it takes less than a page to sketch the problem.

First, a group is nothing but an arbitrary set G, equiped with an operation *, such that:

(a) Closure: for all g and g' in G, g * g' also belongs to G.
(b) Associativity: for all g, g' and g" in G, (g * g') * g" = g * (g' * g").
(c) Neutral element: there is an element e in G, such that
   g = e * g = g * e.
(d) Inverse element: for every g in G there is an element g⁻¹, such that
   g * g⁻¹ = g⁻¹ * g = e.

Secondly, a group is finite if G is a finite set.

Thirdly, we must define what a subgroup is. H is a subgroup of G, if H is a subset of G and H has the group properties for *. A trivial subgroup is the group consisting of the neutral element only, {e}. It is easy to check that the four properties hold.

Fourthly, a group is simple, if its only subgroups are {e} and G itself.

The challenge now is to classify all finite, simple groups. Often an analogy is drawn with prime numbers. In the same way that all natural numbers are built up from the prime numbers, finite groups can be composed starting with the finite, simple groups. Hence, the importance of the classification theorem.

M: Yes, yes, I agree, but what's the problem? If a group of mathematicians tells me that together they have checked all parts of the proof, that is fine with me. I think we already had that discussion before.

P: Exactly, so the same conclusions hold, do they not?

M: Sure, but, once more, what's it to me?

P: With Fermat and the finite, simple groups, as I said before, most likely nothing. But there is something more and that brings me to my third example, namely, the use of computers in mathematical practice. Let me give an example first, the famous four-colour theorem. Part of the proof consisted of checking one by one a fairly large set of cases. For a human mathematician, this would be incredibly tedious and endless work with an extremely high probability that a mistake is made somewhere. So, this task was entrusted to a computer who did the calculations. The final answer of the machine was "OK" and so the theorem was considered proved.

INTERMEZZO

The four-colour theorem states that any planar map divided up into regions can be coloured by four colours in such a way that neighbouring regions get different colours, where 'neighbouring' means that at least a segment and not, e.g., one point is common to both. The problem was formulated for the first time in 1852 by Francis Guthrie and was finally proved (?) in 1976 by Kenneth Appel and Wolfgang Haken (see Wilson [2003] for details). Usually, the name of John Koch is not mentioned, although his contribution is as important as that of the two mathematicians, for Koch provided the algorithms for the computerpart of the proof. Appel and Haken showed that all possible, i.e., infinitely many maps could be reduced to a finite set in such a way that, if all members of this finite set could be coloured with four colours, then all maps can be so coloured.[12] Checking this finite set was done by computer. Although the proof has been simplified and reduced in terms of cases to be examined, the computerpart of the proof remains essential.

M: But are they not trying to reduce this part of the proof?
P: Indeed, they are, but some mathematicians consider this not to be necessary. The proof as we have it today, including the computer part, is fine for them, so one need not bother anymore. Additionally, they draw attention to the fact that such proofs will only become more frequent as, given more complex problems, it will be more likely that parts will become too complex for humans and hence will be left to computers. You see, what I am trying to say, is that in this case mathematicians among themselves started to discuss the problem whether this should count as proof or not. Or,

let me put it differently, your practice is changing, you know it is changing, but you keep sticking to the ideal. You know, I am almost tempted to say, but in a quite positive state of mind: "Get real!"

M: Look, I am a mathematician and what I want to do is mathematics. Does that bother you, for that is the impression I am getting here. Do you want me to quit my job and make myself useful in society? Or what?

P: No, no, no, of course not! Please, keep on doing mathematics, for otherwise I have no longer a topic to think about. When I spoke about a division of labour, this does not mean that we should go separate ways. To a certain extent we do, of course: you are working in a mathematics department in a subfield of mathematics and I am in a philosophy department in the subfield of logic, philosophy of mathematics, and philosophy of science. We have largely separate networks, journals, conferences, you name it. So the division is already there. But that does not imply that we cannot talk *about* one another. What we philosophers are trying to do, is to understand better what mathematicians are doing. For a twofold purpose: a philosophical one, where we wish to contribute to the on-going discussion about the nature of mathematical knowledge, and a mathematical one, where the hope is that mathematicians themselves might be interested in the results of our research. It also means at the same time that there is no need whatsoever to be "afraid" or annoyed when a philosopher says something about what you are doing.

M: As if! I have never been afraid of you, why should I? I know what I am doing...

P: (Interrupts) Exactly!! That is what I am saying: you know what you are doing, but not necessarily *about* what you are doing. See, we can complement each other here.

M: Let me think about that. I have to confess that it sounds a bit funny. Has this ever been done before? I cannot remember ever to have heard about this complementarity thing.

P: Don't laugh, but the best answer is "yes" and "no." What has been done before is that mathematics and philosophy have interacted with one another, but for the obvious reason that the same person practiced both. Think of the Greek philosophers, think of Pascal, Descartes, Leibniz, Newton,...so the answer is "yes," but it also "no" when we are talking about mathematics and philosophy as separate domains. Come to think of it, this is also a question that should be answered by historians, sociologists and philosophers of science and mathematics so that interaction is required as well.

M: Make love, not war, then?

P: Fine with me, better to discuss various ways of love making, then to invent better fighting equipment, I guess.

Here the dialogue between Paul and Martha ends. Next time they met, Martha wanted to discuss with Paul the idea of a mathematical experiment, whether that notion made sense or not, because one of her colleagues claimed to have done such an "experiment" and Paul wanted to find out from Martha how mathematicians decide what problems are interesting or not, i.e., worth the trouble to investigate. After that, they decided to meet on a regular basis.

CONCLUSION

It is obvious that this conversation has taken place in some at present non-existent utopian place in some far (?) away future, where mathematicians and philosophers come to understand the fruitful ways wherein they can complement each other. Because that is the main thesis of this paper: a philosopher of mathematics, by extension a philosopher of science, need not be a "threat" for the mathematician or, by extension, the scientist. True, if all goes well, it should have some impact. If philosophers come up with some well-founded criticisms about mathematical (scientific) practice, then, yes, it should make mathematicians (scientists) think. That being said, it should be equally clear that the impact is symmetrical: the philosopher too should take into account what the mathematician or scientist is doing when they come up with some interesting ideas, which, as it happens, they frequently do. To avoid wars and instead to opt for working together, must surely be considered as a form of progress beyond C. P. Snow's two cultures.

NOTES

1. The paper contains such beautiful passages as this: "the π of Euclid and the G of Newton, formerly thought to be constant and universal, are now perceived in their ineluctable historicity; and the putative observer becomes fatally de-centered, disconnected from any epistemic link to a space-time point that can no longer be defined by geometry alone" (p. 222).

2. It is, e.g., quite unclear why a philosopher such as Régis Debray wants to use Gödel's theorems to illustrate certain claims about human society, such as the claim that a society can never be "closed," in the sense that it would have perfect control over itself. True, there are elements present here that refer to the

object-level including the meta-level, to (vicious) circularity, to self-reflection, and so on, but that is still a long way from what Gödel has shown. Chapter 10 of Sokal & Bricmont (1997) lists some more examples about use and abuse of Gödel's theorems. However, if one is really interested in that theme, rather than Sokal & Bricmont (1997), read Franzén (2005).

3. Especially the French philosophers are the victims of their attacks: Jacques Lacan, Julia Kristeva, Bruno Latour, Jean Baudrillard, Gilles Deleuze, Félix Guattari, to name some of them.

4. At least in that respect many a philosopher does not mind being compared to Socrates.

5. And, for that matter, mathematicians too as is so often claimed.

6. Although at a later stage, they did get partially involved, see, e.g., Koertge (1998).

7. There is a curious story attached to my choice of names. One might think that I deliberately picked out "Paul" as this is meant to be a contribution to a Festschrift, but it is a bit more complicated. I have been working on this paper for some time now, not yet with the intention of having it published. So, before I was invited to contribute to this volume, my imaginary Paul was already in existence. When the invitation came, this seemed to good to be true, for, unless I am gravely mistaken, some of the views of my imaginary Paul do correspond with the views of the "real" Paul. Perhaps I did have him unconsciously at the back of my mind. In this sense, although the paper does not deal directly with constructivism, radical or otherwise, or with educational matters, it is a hommage to Paul Ernest.

8. No story attached to this name, sorry!

9. See Lakatos (1976), pp. 106–126. Actually, Lakatos pays considerable attention to this topic of translation from geometry into more abstract spaces. Without going into the details but with the purpose of letting the reader "taste" the difference, this a a modern version of Euler's theorem (actually, it can be derived from it): if z is an element of Z_F and $z = a_1e_1 + \ldots + a_{n-m+1}e_{n-m+1} + b_{n-m}e_{n-m} + \ldots b_ne_n$, where $e_1, \ldots, e_{n-m+1}$ are not in T and $e_{n-m}, \ldots, e_n$ are in T, then $z = a_1A_fe_1 + \ldots + a_{n-m+1}A_fe_{n-m+1}$. Cubes, pyramids, and other geometrical objects have, apparently, completely disappeared.

10. Sidney Harris is an American cartoonist whose work focuses almost exclusively on science and technology. The quote is also the title of one of his collections.

11. If P(x) stands for "x is a proof," M(x) for "x is a mathematician" and U(x,y) for " x understands y," then, for the die-hards among us, the statement is the following: $(\exists x)(P(x)\ \&\ \sim(\exists y)(M(y)\ \&\ U(y,x)))$. This statement is obviously not contradictory with the statement that a group of mathematicians can understand the proof.

12. A more correct phrasing of that sentence is that one actually searches for counterexamples and has to show that none of the possible candidates is indeed such a counterexample. It is therefore not a matter of actually coloring a large, though finite set of actual maps. If anything, it is much closer to pattern recognition.

REFERENCES

Devlin, K. (1988). *Mathematics: The new golden age.* Harmondsworth, UK: Penguin.

Franzèn, T. (2005). *Gödel's Theorem. An incomplete guide to its use and abuse.* Wellesley, MA: A. K. Peters.

Gross, P. R., & Levitt, N. (1994). *Higher superstition: The academic left and its quarrels with science.* Baltimore, MD: Johns Hopkins University Press.

Koertge, N. (Ed.) (1998). *A house built on sand. Exposing postmodernist myths about science.* Oxford: Oxford University Press.

Kuhn, T. (1977). *The essential tension. Selected studies in scientific tradition and change.* Chicago University Press.

Lakatos, I. (1976). *Proofs and refutations. The logic of mathematical discovery.* Cambridge: Cambridge University Press.

Sokal, A. (1996a). Transgressing the boundaries: Towards a transformative hermeneutics of quantum gravity. *Social Text,* nrs., *46/47,* 217–252.

Sokal, A. (1996b). A physicist experiments with cultural studies. *Lingua Franca, May/June,* 62–64.

Sokal, A., & Bricmont, J. (1997). *Impostures intellectuelles.* Paris : Odile Jacob.

Wilson, R. (2003). *Four colors suffice. How the map problem was solved.* NJ: Princeton University Press.

CHAPTER 5

THE ANALYTIC/SYNTHETIC DISTINCTION IN KANT AND BOLZANO

Michael Otte
Bielefeld, Germany

INTRODUCTION

No education, no mathematics education in particular without a philosophy and an epistemology. Philosophical reflection, when taken in its broader concerns, rather than as the activities of a specialized group of academics, is essentially concerned with the universality of human being and; therefore, with the relation between the particular and the universal. Plato had been the first to state this in clear terms and he had used mathematics to illustrate the issue epistemologically. As the ideal and essential can only be itself, rather than its opposite, differently from what happens in case of the particular and concrete, we can build up consistent knowledge about the universal, but not about the empirically concrete, which the *Sophists* have shown to remain always ambiguous. This provides mathematics with a privileged role. Mathematics gives us an example of how to answer a given question by building up a consistent theory (see for example *Phaedo*).

Relatively and Philosophically Eᵃrnest, pages 39–55
Copyright © 2009 by Information Age Publishing

39

It is not by chance that a Platonist like Leibniz introduced our notion of formal proof by contradiction into mathematical reasoning again. Leibniz did, in fact, not clearly distinguish between logical an factual truth (Neemann 1981, 76). Since the 19th century only and with the creation of pure mathematics, mathematicians have begun to appreciate this view of mathematics as the science the possibility, basing it on the principle of non-contradiction. But, in the meantime the problem of generalization from the particular or the problem of induction, as it is also called, had appeared on the scene, because rationalism could no more be based in God but had to take subjective or social concerns into account.

Either one acknowledges the synthetic character of generalization or one considers all general affirmations as mere postulates or hypotheses. In the first case mathematics becomes synthetic, in the second analytic. As soon as one formulates the concept of arithmetical sum, for instance, in terms of the cardinality of sets (intuitively assuming the existence of the latter and generalizing from this intuition), the concept is obtained as a law, and the arithmetical theorems in question thus become synthetical (Cassirer 1907). As soon as the natural numbers, however, are introduced axiomatically, introducing the concept of sum recursively on the basis of the successor operation of ordinal numbers, arithmetical theorems become analytic, at least upon the dominant linguistic understanding of logic and axiomatic. If one conceived, however, of logic in metaphysical terms, rather than as merely linguistic arithmetic might be considered synthetic. Kant, as well as of Bolzano considered arithmetic to be throughout synthetic. But let us look more closely at the matter, because things seem to be more involved.

THE COMPLEMENTARITY OF EXTENSIONAL AND INTENSIONAL APPROACHES TO PROOF

The logical empiricists conceived of logic in linguistic terms and of knowledge as a combination of tautological formal structure, on the one side and piecemeal empirical information, on the other side. "Pure logic and atomic facts", says Russell, "are the two poles, the wholly *a priori* and the wholly empirical" (Russell, 1914, p. 63). A first appearance of this approach can be traced back to Leibniz, Grassmann and Boole and to Boolean calculus. Russell describes how he had become acquainted with these developments:

> The modern development of mathematical logic dates from Boole's *Laws of Thought* (1854). But in him and his successors, before Peano and Frege, the only thing really achieved, apart from certain details, was the invention of a mathematical symbolism for deducing consequences from the premises which the newer methods shared with those of Aristotle. This subject has considerable interest as an independent branch of mathematics, but it has very little

to do with real logic. The first serious advance in real logic since the times of the Greek was made independently by Peano and Frege—both mathematicians. They both arrived at their logical results by an analysis of mathematics. Traditional logic regarded the two propositions, *Socrates is mortal* and *All men are mortal*, as being of the same form; Peano and Frege showed that they are utterly different in form. (Russell, 1914, p. 49/50)

Russell criticizes that logicians had ignored the distinction of type theory, according to which the two propositions cited are of different logical type. Stated in set theoretical terms, these logicians had failed to distinguish a set from the totality of its elements—collection-as-one vs. collection—as many—that is, they had ignored the difference between the subset-set relationship and the element relationship.

Why are the two sentences *Socrates is mortal* and *All men are mortal*, so different as Russell emphasizes? Now, first of all one should mention their very different truth conditions. The mortality of Socrates might be observed directly, while the mortality of humans in general cannot be deduced directly from empirical facts. "If we would know each individual man and know that he was mortal, that would not enable us to know that all men are mortal, unless we knew that those were all the men there are, which is a general proposition" (Russell, 1914, p. 65). This general proposition implies a hypothesis about the existence of objects of the world. The relation between the statements *A*: "x is a human being" and *B*: "x is mortal" is a relation of material implication, that is, it only says that it cannot be that *A* is true and *B* false.

One might provide a different proof of the mortality of Socrates by modus ponens: "All men are mortal. Socrates is a man. Therefore Socrates is mortal". The particular is explained by the general. The proof is based on the belief that it belongs to the nature of humans to be mortal. There seems thus a necessary connection between humanity and mortality. Aristotelian demonstrative science concerns itself exclusively with necessary knowledge and deals with the universal, not the particular and contingent; it treats individuals, not in their own right, but as falling under universals. We have knowledge of a particular thing, according to Plato or Aristotle, when we know what it is (*Metaphysics*. B 996b). This means that the universal determines the particular. This makes, however, all knowledge analytic, negating any distinction between our inner and outer world. To classical ontologism all true propositions had been analytically true.

There is a third way to see such a proof of Socrates mortality. One does not, like Russell infer from the mortality of all individual men there are in the universe the general proposition, but one assumes that for each human being there is a necessary connection between the fact of being human and the fact of being mortal and generalizes from this, supposing then that there should be a necessary connection between humanity and mortality.

This kind of logical implication is stronger than material implication, because it derives not merely from the truth values of *A* and *B*, but from its content. To reduce logical relation to material implication one must assume something about the objects that exist in reality or that could exist, as Russell had pointed out.

Logical implication prevails in axiomatic. We do not assume anything about the world, when we set up a system of axioms but define merely some concepts whose intensions consist in the set of logical implications of the axioms. A concept is to be defined, as Moritz Schlick said with respect to Hilbert's axiomatization of geometry, by the fact that certain conclusions can be drawn about it. And Hilbert had explained his axiomatic approach along these lines. He writes:

> A + 1 = 1 + A, where A is a number sign, cannot be negated from a finitist point of view…because this equation does not stand for an infinity of particular numerical equations, connected by the word *and,* but represents a hypothetical judgment, which affirms something in case a number sign is given. (Hilbert 1964, p. 91)

To explain a statement like 2 + 2 = 4, for example, one first argues, as in empirical discourse, that this proposition expresses a simple matter of fact, do be easily verified if we accept ostension. After a while one goes on to try and give an *explanation* of this fact. This endeavor implies a change of perspective, a jump to a level of different logical or categorical type. The law gives a unified account of what is otherwise a mere series. In this endeavor one uses the general and ideal or abstract to explain the particular and concrete, or seemingly concrete, in exactly the same manner in which Newton's laws are used to explain simple mechanical phenomena, or Ohm's law is used to explain the facts of electricity. The general, as used in scientific explanations, the associative law of algebra, in the present case of 2 + 2 = 4 for instance, is less sure from an empirical point of view than the individual facts to be founded on it. The less certain is used to explain the more certain, because what could be more certain than a particular fact.

Such a strategy makes sense if it is employed exploratively and predictively, even though the predictions made can never be absolutely sure. The objectivity of such an approach lies in the intended applications. Mathematical axioms in the sense of Hilbert are not fundamental and self evident truths, but are hypotheses which have to prove their fertility and objective value by the intended applications they enable. They are not knowledge, but generate knowledge and orient action.

What matters in the end are the relations rather than the relata. For example, one may encounter two very different types of theories of ethics in the history of philosophy. One is based on a conception of man and his nature or his essential characteristics while the other concentrates on the motives and causes of human behavior hypostatizing predicates, like when one says "human industry is laudable," rather than saying "the industrious man is laudable" or "greed is abominable."

There are two main senses of industry, for example, (a) the human quality of sustained effort and application, (b) an institution or set of institutions for production or trade. One has observed that only from the 18th century onwards the sense of industry as an institution began to come through again (Raymond Williams *Keywords,* 1975, p. 165). So, relational thinking implies the idealization of relations and predicates and reasoning about them.

We encounter now a sort of semiotically modified, that is, representational Platonism. Charles Peirce (1839–1914), for example, defines a sign "as something which stands to somebody for something in some respect or capacity. It addresses somebody, that is, creates in the mind of that person an equivalent sign, or perhaps a more developed sign. That sign which it creates I call the interpretant of the first sign. The sign stands for something, its object. It stands for that object, not in all respects, but in reference to a sort of idea, which I have sometimes called the ground of the representamen. "Idea" is here to be understood in a sort of Platonic sense, very familiar in everyday talk" (CP 2.228; CP 2.275).

While the classical formula portrays the sign in terms of a dyadic relationship, the Peircean definition conceives of it in terms of a triadic structure, the third element, besides object and interpretant being an idea or hypostatic abstraction. Especially mathematical reasoning in its more involved cases implies generalization, that is, the introduction of new ideal objects, which are a result of hypostatic abstraction which then are used to further progress (see Peirce: CP 4.234, 4.235, 4.463, 4.549, 5.447, 5.534).

Hypostatical abstraction is prominent in mathematical thinking from its very beginning and hence from results Russell's concern. We count 1, 2, 3,... and then jump on to a different categorical level making statements about the set of natural numbers in general. Again one should emphasize that the axiom of complete induction, which we use in reasoning about the natural numbers in general is a mere postulate and is essential to establish standard arithmetic. Or think of the moving point generating a line and the line in motion generating the plane. Grassmann's brilliant ideas about spaces of arbitrary dimension, which marked an essential breakthrough over classical "intuitionism" are a result of such imagination. Intuition is directed at the hypostatical ideas abstracted from activity.

For example there had always been the problem of the "general triangle" in geometry taken the fact that a diagram seems to be just a particular design. Already in 1710, Berkeley had asked the readers of Locke's *Essay concerning Human Understanding* to try to find out whether they could possibly have "an idea that shall correspond with the description here given of the general idea of a triangle." And to this logical impossibility he proposed a "representational" solution, saying that "we shall acknowledge, that an idea, which considered in itself is particular, becomes general, by being made to represent or stand for all other particular ideas of the same sort" (Berkeley 1975, p. 70). Jesseph has characterized Berkeley's philosophy of geometry by the term "representative generalization" and he writes: "The most fundamental aspect of Berkeley's alternative is the claim that we can make one idea go proxy for many others by treating it as a representative of a kind" (Jesseph 1993, p. 33).

On such an account a general triangle is a free variable, like the terms in axiomatic descriptions, and not a collection of determinate triangles. It is an idea, which governs and produces its particular representations. And which properties are essential to a "general triangle," depends on context, on the activity and its goals. If the task, for instance, is to prove the theorem that the medians of a triangle intersect in one point, the triangle on which the proof is to be based can be assumed to be equilateral, without loss of generality—because the theorem in case is a theorem of affine geometry and any triangle is equivalent to an equilateral triangle under affine transformations. This fact considerably facilitates conducting the proof because of such a triangle's high symmetry. And the fact that we are able to prove theorems about a triangle in general shows that that it is real, says Peirce (CP 5.181).

We may notice here a complementary aspect of relational thinking, that is, its instrumental character.

Even the diagrams of Euclid could be interpreted in two complementary ways. Ian Mueller, for example has described the situation in relation to Euclid's diagrammatical proofs as follows:

> The Euclidean derivation is a thought experiment...the major obstacle to an acceptance of the interpretation of Euclid's arguments as thought experiments is the belief that such arguments cannot be conclusive proofs. In particular, one might ask how consideration of a single object can establish a general assertion about all objects of a given kind. Part of the difficulty is due, I think, to failure to distinguish two ways of interpreting general statements like "All isosceles triangles have their bases angles equal." Under one interpretation the statement refers to a definite totality...and it says something about each one of them. Under the other interpretation no such definite totality is presupposed, and the sentence has much more conditional character—"If a triangle is isosceles, its two base angles are equal." A person who interprets a generalization in the second way may hold that the phrase "the class of isosceles triangles" is meaningless because the number of isosceles triangles is absolutely indeterminate. (Mueller, 1969, p. 292, 299–300)

Mathematics then reasons starting from the meanings of certain representations, rather than from supposed characteristics of a class of objects.

Theoretical concepts on such accounts are not empirical abstractions but are operative schemata, like in modern axiomatics in the sense of Hilbert. Mathematics becomes intensional like logic and it must be complemented by some intended applications.

This complementarity of relational and referential thinking appears in all mathematics because all statements require a subject about which something is stated. Even in pure mathematics indexical signs are therefore indispensable. One might think, Peirce himself says:

> That there would be no use for indices in pure mathematics, dealing, as it does, with ideal creations, without regard to whether they are anywhere realized or not. But the imaginary constructions of the mathematician, and even dreams, are so far approximate to reality as to have a certain degree of fixity, in consequence of which they can be recognized and identified as individuals. (Peirce, CP 2.305)

The indices occurring in pure mathematics refer to entities or objects that belong to models, rather than to "the real world." Mathematics become characterized by a complementarity of deductive proof oriented and model theoretical thinking.

Modern epistemology does no more ask, what is the epistemic subject, what is an object of knowledge, that is, it does not try and explain the relationship from analyzing the nature of its *relata*, but concentrates on the relations from the very beginning. These relations are defined in terms of intuitions, concepts, languages, practices, instruments, institutions, and so on.

For instance, one group of logical philosophers emphasizes the universality of language and hence has to accept the fact that one cannot in language point to an object without describing it. Another assumes non-linguistic relationships to objective reality as given and may then draw distinctions without describing or justifying them, distinctions without a reason, so to say.

A comparison of Kant and Bolzano and of their respective philosophies of science and mathematics proves illustrative.

Kant

It was Kant who provided the analytic-synthetic distinction with fundamental importance to epistemology. The analytic/synthetic distinction lies at the very heart of Kant's critical philosophy. Kant had realized that human practice and activity make up the essence of reality and from this derived the importance of the analytic/synthetic distinction. In his *Critique of Pure Reason* Kant formulated the problem as the question "How are synthetic judgments a priori possible?" (Wie sind synthetische Saetze a priori moeglich?"—Prolegomena §5; see also reply to Eberhard: *Werke*, vol. V,

364–367). Kant gives the following description of the analytic/synthetic distinction:

> In all judgments wherein the relation of the subject to the predicate is thought this relation is possible in two different ways. Either the predicate B belongs to the subject A, as something, which is contained (covertly) in this concept A; or B lies completely outside of the concept A, although it stands in connection with it. In the first instance, I term the judgment analytical, in the second synthetical. (B 11)

In his *Prolegomena* of 1783 Kant provides still another description in terms of a distinction between amplifying (erweiternd) and illustrative (erläuternd) arguments (Kant, 4.266, §2). Thus all generalization depends on synthetic judgments.

Kant's "Copernican Revolution" of epistemology, which opened the doors to nearly all modern philosophy is, as Kant in his controversy with Eberhard had stated emphatically himself, essentially linked to the analytic/synthetic distinction and in particular to the existence of the synthetic *a priori*, because the latter marks the scope and limits of human knowledge:

> We are actually in possession of *a priori* synthetic cognitions, as is proved by the existence of the principles of understanding, which anticipate experience. If anyone cannot comprehend the possibility of these principles...he cannot on this account declare them to be impossible...He can only say: If we perceived their origin and their authenticity we should be able to determine the extent and limits of reason. (KrV, B 790)

Mathematics plays an illustrative role in this endeavour to trace out the limits of human knowledge. To Kant the case of mathematics was so important because mathematics seems, like logic to provide secure and necessarily true knowledge, but differently from logic, "in which reason has only to deal with itself," mathematics has to generalize in order to gain new knowledge and all true generalizations are based on synthetical judgments. To Kant's critical philosophy the insight is essential that mathematics, although necessary and a priori, produces real discoveries and gains new knowledge.

And in fact:

> Mathematics affords us a brilliant example, how far, independently of all experience, we may carry our *a priori* knowledge. It is true that mathematics occupies itself with objects and cognitions only in so far as they can be represented in intuition. But this circumstance is easily overlooked, because the said intuition can itself be given *a priori* and therefore is hardly to be distinguished from a mere pure conception. (KrV, B 8)

Mathematics is not just about meanings but is extensional investigating relationships between extensive magnitudes. "The reason why mathematical cognition can relate only to quantity is to found in its form alone. For it is the conception of quantity only that is capable of being constructed, that is, presented a priori in intuition" (*KrV*, B 742). We cannot know in advance the true essence of things, but have to start from mere perception and appearance. Leibniz God had solved this problem through an infinite analysis of things. Kant abandoned the Gods eyes perspective and he was interested in how we humans come to know.

It is well-known, Kant says, that all our knowledge refers to appearances, and not to the world of "things in themselves" which are not cognizable. Now appearances, Kant continues, "are continuous magnitudes, alike in their intuition, as extensive, and in their mere perception (sensation, and with it reality) as intensive" (KrV, B 212). Mathematics deals with extensive magnitudes and from this results its method, which Kant calls the "construction of concepts." Kant writes: "The construction of a conception is the presentation a priori of the intuition which corresponds to the conception" (B 741). By these means something, which might have implicitly and intuitively been thought about, because it had so far not been included into the conception of the object of thought, now becomes transformed into explicit knowledge.

Kant, accepting Descartes spatial characterization of material bodies, considers the judgment "All bodies are extended" to be analytic, whereas the judgment "All bodies are heavy" is seen as synthetic, because, according to Kant, "the predicate is something totally different from what I think in the mere conception of a body" (*KrV*, B 11).

The analytic/synthetic distinction seems thus to depend on how we see things and how we think them. But one should note here that it is not a matter of free and arbitrary choice how we grasp things. The view that bodies are essentially extended results from Kant's "axioms of intuition" (*KrV*, B 202), which demand that "all intuitions are extensive quantities", such that extension belongs essentially to the notion of a body, but weight does not. Every sensible object is by definition extended and this claim is justified by the way humans gain objective knowledge.

The continuity of space as such:

Does not provide knowledge.... To cognize something in space, a straight line, for instance, I have to draw it and thus to produce synthetically a determined conjunction of the given manifold...and by these means alone is an object recognized. (*KrV*, B 137)

The synthetic unity of consciousness, according to Kant, is "an objective condition of all knowledge. . . . For in the absence of this synthesis, the manifold would not be united in one consciousness(*KrV,* B 138).

Bolzano

Bolzano praised Kant for his "deep insights" into the question of the analytic-synthetic distinction and its importance. It was Kant who, although not having discovered this distinction, "has provided it with the appropriate attention." "And in order to recognize that there are characteristics of an object . . . that nevertheless are not presented by the concept of that object, it is only required to see adequately that distinction" (between analytic and synthetic truths). Kant had claimed "that all theorems of mathematics, physics etc are synthetic truths. He who understands this will also understand that there are innumerous characteristics of an object which can be deduced from the concept of that object, although we do not think them as components of that concept" (WL § 65, 288f).

"However," Bolzano complains, "while there are many followers of Kant's distinction, there are few who have since then properly distinguished between components (Bestandteilen) of the concept and characteristics (Merkmalen) of the object" (loc.cit, 289). Even Kant himself had not properly observed this difference (loc.cit, 292). As Kant had identified the concept with a set of some properties of its object, rather than as a representation of that object, his distinction between essential and other characteristics seemed somewhat ambiguous.

Now Bolzano's program is a foundationalist program; he is not interested in the so called context of discovery. Bolzano accordingly, criticized Kant saying that the latter had confounded mathematics as such with the ways we humans might come to know it. Bolzano follows Leibniz views in this respect (Nouveaux Essais IV).

Secondly Bolzano's notion of analyticity relates, as we shall see, to the specific representations of knowledge claims, rather than to knowledge as such. It is thus freed from difficulties of Kant's or Leibniz' distinction between essential and additional attributes of a subject and also from problems with possible restrictions of concept formation. Bolzano was concerned with what we say and its logical rules and he saw that consequently the foundations of a philosophy of science must be searched for in semantics. His monumental *Wissenschaftslehre* (Doctrine of Science) established Bolzano, in fact, "as the founder of the semantic tradition" (Coffa 1991, p. 23).

The main elements of Bolzano's *Wissenschaftslehre* that are to be taken into account in order to understand his approach to analyticity are the notion *Vorstellung an sich* (representation in itself) and *Satz an sich* (proposition in itself). Bolzano defines a representation in itself (*Vorstellung an sich*) negatively by asking the reader to take the notion of representation in the common sense and then abstract from everything which is subjective or accidental about it. All of Bolzano's declarations of the fundamental notions of the *Wissenschaftslehre* proceed in this manner. The fundamental property of a sentence is its truth or falsehood. And as a sentence can only be true if its subject represents something, Bolzano had to assume a very differentiated ontology. The sentence "Unicorns exist" is true, for example, when understood in the context of fairy tales.

An objective representation in Bolzano's sense is to be distinguished from its object, from that which it represents. We must thus distinguish between three different things, the representation as such, its concrete expression or form or its realization in thought and its objective content. The object of the representation may or may not exist. The square root of −1, for example, is a representation without an object (WL §49). However, "when uttering the proposition *The representation of the square root of −1 is compound*, then the representation of the subject of that proposition is an objective representation, its object being $\sqrt{-1}$; this $\sqrt{-1}$ as such, however, being a part of the representation, does itself not represent an object" (GL §4; see also WL §71). A representation thought of or expressed, is always about something, namely about the representation as such or about the meaning as hypostatized as an object of thought or communication. And anything can become an object of thought, such that there are no limitations in the Kantian sense of sensible experience (WL §81).

In the same manner a proposition in itself is to be distinguished from a proposition expressed in words or acted upon in thought. Bolzano's *Wissenschaftslehre* is the set of propositions in themselves, or rather a presentation of their general properties and their hierarchical order. A proposition is either true or false; and this permanently so.

That certain propositions, like "this flower smells pleasant," or "a bottle of wine costs 10 thaler" appear as sometimes true and sometimes false, depending on circumstances, is due to our disregarding that the proposition in question does not remain the same. "This," for example, is an indexical sign with different referents depending on context. And in the second example we assume, says Bolzano, tacitly that there is a context of time and space when we hear somebody making such a judgment (WL §147) and the proposition therefore does not remain the same.

So we really assume, without being aware of it, certain representations in a meaningful or valid proposition as variable or varying. If this is so then it seems worthwhile, says Bolzano, to investigate explicitly how propositions behave upon the variation of certain of their representations. "If we consider not only the fact of the truth or falsehood of a proposition but also how all the propositions, which can be derived from it by varying certain of its representations are related to the question of truth and falsehood, then this will lead us to discover some remarkable properties of propositions" (WL §147).

A judgment "contains a proposition, which is either in accordance with the truth or not; in the first case we call the judgment correct and in the second case incorrect" (WL §34). What seems in particular important about certain propositions, Bolzano writes, is "that their truth or falsehood does not depend on the individual representations that constitute them, but remains the same if one changes the latter, taken for granted that the proposition stays objective" that is, meaningful.(WL §148).

Hence Bolzano's definition of analyticity:

> If there is a single representation (eine einzige Vorstellung) in a proposition which can be arbitrarily varied without disturbing its truth or falsity... then this character of the proposition is sufficiently remarkable to distinguish it from all others. I permit myself thence to call propositions of this kind, borrowing an expression from Kant, *analytic*, all others, however, *synthetic* propositions. (WL §148)

This definition corresponds to Bolzano's model theoretical conception of deducibility. Bolzano defines deducibility, in fact, by saying that whenever a substitution of a component in the antecedents makes them true, the same substitution will make the conclusions true (WL §155).

The arithmetical propositions are exemplary synthetic propositions to Bolzano, while the language of algebra provides examples of analytic propositions. Algebra is, however, not conceived of in syntactical terms by Bolzano. It is a kind of passing from the particular to the general and universal. Generalization means to introduce variables, rather than abstraction from content. "Every algebra has an arithmetic" (Spencer-Brown), or, in Bolzano's terms, every representation represents something. Or, finally as J. Proust had stated it: "The refusal of formal syntax originates in a conception of the proposition as the only true object of the logician" (Proust 1989, p. 58).

Bolzano's definition of analytical judgments reflects this clearly. Bolzano's semantic view of logic is also manifest in his notion of deducibility, which is closely connected with his definition of analyticity. I say, writes Bolzano, "that the propositions M, N, O, ... are deducible from the sentences A, B, C, D, ... with respect to the variables i, j, ... if every set of representations which, put at the places i, j, ... making the A, B, C, ... true, will also make the M, N, O, ... true" (WL §155). This definition again suggests an interpretation in terms of models.

Finally, all modal forms have, according to Bolzano, to be related to the notion of Being (*Sein*).

What concerns the notion of necessity, in particular, we say in fact, I believe, that the being of a certain object A is necessary only if there is a purely conceptual truth of the form: B is, where B is a representation that contains A. We say that God necessary is, because the proposition, that God is, is a purely conceptual truth. (WL §182)

Bolzano warns that it is not always easy to see if a sentence is analytic or synthetic. For example, the proposition, "An erudite man is a man" is not analytic, because it simply wants to say that "even an erudite man is fallible," whereas the other: "Each effect has a cause" is analytic because it is of the form "A, which is B, is B" (WL §148). The paradigmatic analytical proposition is of the form "A has A," like "this triangle has 3 vertices." Bolzano calls these judgments "identical" (WL §148). This means that Bolzano distinguishes analyticity in a wider and narrower sense and only this narrow sense has been preserved until today.

Even a logical rule like "A implies B and B implies C results in A implies C" is synthetic, according to Bolzano (WL §315), because the representations A, B, C cannot be indeterminate as to all their characteristics, as logicians commonly believe. B must for example be "an idea which can be predicated of all A" (WL §7); etc. etc.

We have always to ask ourselves what a sentence means, that is, we have to find the "Satz an sich" (*SAS*). Thus Bolzano's objective semantics and the Platonic and hierarchically structured universe of objective meanings is essential to his whole conception of explanation.

For instance, Bolzano considers, like Kant, the proposition "The angle sum of a triangle is equal to two right angles" as synthetic because it represents no identity but an equality between co-extensional terms differing in meaning (WL §148). Generalizing this theorem, in the manner mathematicians are accustomed to do, results in the proposition, "The angle sum of a n-gon is equal to $(n-2)$ times two right angles," which is, however, analytic.

But is not "the angle sum of a 3-gon is $(3-2)$ right angles," on the one side, analytic and on the other side equivalent to: "The angle sum of a triangle is equal to two right angles"? It depends. People who consider mathematics to be extensional throughout or who see arithmetic as being analytic would see an inconsistency in Bolzano's definition of analyticity. People, however, like Bolzano, who think that meaning is what constitutes real knowledge and who would refuse to simply identify co-extensive concepts, have no such problem here.

Two concepts A and B are not the same, even if contingently or necessarily all A's are B's and vice versa, because different concepts help to establish different kinds of relationships between subject and object of knowledge and possible knowledge becomes real or actual only by means of such relationships. Two concepts could be extensionally equivalent and yet might function differently, within a cognitive or communicative context. There are also neither two completely equal propositions (Saetze an sich) (WL §150), nor two identical representations (Vorstellungen an sich) (WL §91), says Bolzano.

Bolzano quotes a contemporary author who had written that it depends on the definition of "triangle" whether the proposition "The angle sum of a triangle is equal to two right angles" is synthetic or analytic. Bolzano disagrees because the identity of a proposition (the meaning of a sentence) changes if we substitute co-extensional terms for each other (§ 148, Annot. 4). The subject term of a proposition is merely a manner of presenting the related object, rather than the latter itself. Bolzano emphasizes a principle according to which whenever we wish to say something about a certain thing we have to use not the thing itself, but its name or designation. This seems trivial enough, but nevertheless it very often is not present to the mind. When one says "I see a white azalea" this is not strictly true because I do not see a word or a sentence but an image of something, which the English speaking people have chosen to call an azalea.

"The angle sum of an equilateral triangle is equal to two right angles" is analytic because the attribute "equilateral" is interchangeable (WL § 447). All premises which do not contribute to the truth of the conclusion make the latter analytic. I could not, writes Bolzano, "be satisfied with a completely strict proof if it . . . made use of some fortuitous, alien intermediate concepts."

In the same sense is the sentence "This triangle has an angle sum of two right angles" analytic. Instances or examples of conceptual propositions are analytic, because the subject term can vary over the range of the extension of the concept. So "this flower is a plant" is analytic, whereas "this flower smells pleasant" is synthetic.

Unprovable, or basic propositions must, according to Bolzano, be among those whose subjects and predicates are completely simple concepts, they must be synthetic. Bolzano believed that different cases of one and the same issue should be formulated in terms of different propositions, like in Euclidean geometry. The law of cosine, for instance, in the cases of the acute- respectively obtuse-angled triangles signifies two different cases requiring different arguments. "Euclid was right in formulating two different propositions here," writes Bolzano (Bolzano, 1810/1926, p. 61). And he considers the theorem of Pythagoras as a third independent case. In modern mathematics on might instead, using the cosine function, state just one theorem which would then to be considered analytic. But, continuous variation and analysis comprise the method of discovery, not of proof. Bolzano's views are very similar indeed to those of Euclid or Aristotle.

Different propositions referring to the same truth may one be analytic the other synthetic, as we have noticed in the case of the theorem about the angle sum of a triangle, depending on how they have been proved. Bolzano's "*Wissenschaftslehre*" contains a distinction between proofs that verify, being intended to create conviction or certainty, and others, which "derive the truth to be demonstrated from its objective grounds. Proofs of this kind could be called justifications (*Begründungen*) in difference to the others which merely aim at conviction (*Gewissheit*)" (WL §525). Proofs should be based on necessary conditions and should be "pure."

Bolzano criticised, for example, Gauss's proof of the fundamental theorem of algebra of 1799, because Gauss had at that occasion employed geometrical considerations to prove an algebraic theorem. Bolzano did not, as is often claimed, doubt the validity of Gauss's arguments and he did not question the certainty of our geometrical knowledge, but criticised the "impurity" of Gauss' proof of 1799. Bolzano did not question the certainty of our geometrical knowledge.

Bolzano writes: "It is an intolerable violation of the principles of sound methodology to deduce truths of pure (or general) mathematics (that is of arithmetic, algebra or analysis) from considerations which belong to a part of mathematics, which is merely applied or special mathematics, in particular from geometry. . . . For in fact, if one considers that the proofs of the science should not merely be convincing arguments, but rather justifications, i.e.,—presentations of the objective reason for the truth concerned, then it is self-evident that the strictly scientific proof, or the objective reason of a truth which holds equally for all quantities, whether in space or not, cannot possibly lie in a truth which holds merely for quantities which are in space. On this view it may on the contrary be seen that such a geometrical proof is really circular. For while the geometrical truth to which we refer here is extremely evident, and therefore needs no proof in the sense of confirmation, it nonetheless needs justification" (Bolzano).

Both Kant as well as Bolzano had emphasized the fact that knowledge is conditioned by its means and by the human condition in general. Only to Bolzano communication and the social orientation in general became all important, rather than the certainties of the individual's imagination. His *Wissenschaftslehre* is nothing but a didactical treatise and a methodological investigation about how to write textbooks in the various scientific disciplines.

There is a second cause which led to sensitivity with respect to the subjective and social conditions of knowledge and in consequence to a renewed interest in the analytic/synthetic distinction. Philosophers and scientists or mathematicians searched for means to cope with the yet unknown and insecure, rather than reproducing what was already known or believed. For instance, the principle of non-contradiction became and explorative means to investigate into the logically or mathematically possible and to make predictions.

Mathematics, like logic, must have a specific subject matter, rather than being a mere science of form or structure. Bolzano had long hesitated to be explicit about the subject matter of mathematics, because the traditional definition of mathematics as the science of quantity did not suit him for lack of an appropriate definition of the term "quantity". Nearly at the end of his life he wrote:

> I am not afraid to confess that the definition of the concept of *Quantity*... has caused me more efforts than the explanation of all the other notions of this science (of mathematics) and that I have with respect to no other case changed my mind so often. If one determines mathematics as the science of quantity (a definition to which I have essentially returned now) then one takes the word quantity in a wider sense since one counts the doctrine of numbers among the more important mathematical disciplines. According to this wider sense we understand by *Quantity* everything which is of such a kind that between two exemplars of that kind only two possible relations could persist: they must either be equal or one must be equal to a part of the other. (GL, part I, §1)

It might thus be illustrative to compare Grassmann's and Bozano's notions of equality. Bolzano interprets (mathematical) equations $A = B$ as equality of co-extensional terms, as saying that two representations (Vorstellungen) A and B have the same extension. Bolzano calls such representations "interchangeable representations" (Wechselvorstellungen) (GL part III, §44; WL §91–96; §108). Equality is in contrast to this, according to Grassmann, to be defined in strictly operative terms as substitutivity *salva veritate,* that is, syntactically. Logic is a calculus, according to Grassmann, rather than a universe of meanings. For example, Grassmann considers the geometrical congruence relation not as an equality relation, because it violates the rule that "equals added to equals result in equals". Bolzano, in contrast, interprets this fact as implying that the rule of equals cannot be used to define equality (Otte, 1989). Truths and mathematical truths in particular can only be discovered rather than being constructed. This view led Bolzano to admit a complex ontological furniture.

Now, to Grassmann or Peirce or Hilbert the theory as a whole furnishes the context from which one has to determine questions of meaning. To Bolzano it is propositional language which fulfills this function.

REFERENCES

Bolzano, B. (1974). Beitraege zu einer begruendeteren Darstellung der Mathematik . Hans Wussing (Ed.). Darmstadt : Wiss. Buchges.

Bolzano, B. WL = Wissenschaftslehre : 4 vols. ; Wolfgang Schultz (ed), Aalen.

Bolzano, B. GL = Nachgelassene Schriften Bd. 7 Grössenlehre. [1], ‚Einleitung zur Grössenlehre' und ‚Erste Begriffe der allgemeinen Grössenlehre'—1975, in: Eduard Winter, Jan Berg(eds.), Bernard-Bolzano-Gesamtausgabe–Stuttgart-Bad Cannstatt : Frommann

Cassirer, E. (1907), Kant und die moderne mathematik, *Kant-Studien*

Coffa, J. A. (1991). *The semantic tradition from Kant to Carnap,* Cambridge UP.

Hilbert, D. (1964). Ueber das Unendliche, in: *Hilbertiana,* Darmstadt.

Jesseph, D. M. (1993). *Berkeley's philosophy of mathematics.* Chicago: The University of Chicago Press.

Kant, I., KrV = *Kritik der reinen Vernunft,* Werke, vols. III + IV/ed. Wilhelm Weischedel .–Wiesbaden.

Mueller, I. (1969). Euclid's elements and the axiomatic method. *The British Journal for the Philosophy of Science, 20,* 289–309

Neemann, U. (1981). Die Unterscheidung von logischer und faktischer Wahrheit, *Zeitschr. F. Allg. Wissenschaftstheorie,* XII, 75–97.

Otte, M. (1989). Gleichheit und Gegenstaendlichkeit in der Begruendung der Mathematik im 19. Jahrhundert. In G. Koenig (Ed.), *Konzepte des mathematisch unendlichen im 19. Jahrhundert,* Goettingen

Peirce, Ch. S. CP = *Collected Papers of Charles Sanders Peirce,* Volumes I-VI, ed. by Charles Hartshorne and Paul Weiß, Cambridge, Mass. (Harvard UP) 1931–1935, Volumes VII-VIII, ed. by Arthur W. Burks; Cambridge, Mass. (Harvard UP) 1958 (quoted by no. of volume and paragraph)

Proust, J. (1989), *Questions of form.* Minneapolis: University of Minnesota Press.

Russell, B. (1914). *Our knowledge of the external world.* London : Allen & Unwin.

Williams, R. (1975). *Keywords,* London: Penguin.

CHAPTER 6

AESTHETICS AND CREATIVITY

An Exploration of the Relationships between the Constructs

Astrid Brinkmann
University of Muenster, Germany

Bharath Sriraman
University of Montana

In honor of Paul Ernest's 65th Birthday

ABSTRACT

In this contribution, we report on an ongoing study that examines the relationship between aesthetics and creativity among working mathematicians. Writings of eminent individuals indicate that aesthetics is an important component of mathematical creativity, however we were interested in researching this relationship among the normal working mathematician. Anecdotally speaking, many working mathematicians often convey a reciprocal relationship between aesthetics and creativity, particularly when mathematical results and proofs are arrived at with considerable strain and stamina. We report on the findings of our ongoing study among working mathematicians in the

Relatively and Philosophically E^arnest, pages 57–80

U.S.A and Germany, and make a case for emphasizing the aesthetic dimension in mathematics education.

INTRODUCTION

In the general literature one finds several reports concerning the use of aesthetics as a guide when formulating a scientific theory, or selecting ideas for mathematical proofs.

The first who introduced mathematical beauty as well as simplicity as criteria for a physical theory was Copernicus (Chandrasekhar, 1973, p. 30). Since then, these criteria have continued to play an extremely important role in developing scientific theories (Chandrasekhar, 1973, p. 30; Chandrasekhar, 1979, 1987). This is especially so for truly, creative work that seems to be guided by aesthetic feeling rather than by any explicit intellectual process (Ghiselin, 1952, p. 20). Dirac (1977, p. 136), for example, tells about Schrödinger and himself:

> It was a sort of act of faith with us that any questions which describe fundamental laws of nature must have great mathematical beauty in them. It was a very profitable religion to hold and can be considered as the basis of much of our success.

Van der Waerden (1953) reports that Poincaré and Hadamard pointed out the role of aesthetic feeling when choosing fruitful combinations in a mathematical solution process. More precisely, Poincaré asked how the unconscious could find the "right" or fruitful, combinations among the many possible ones. He gave the answer: "by the sense of beauty, we prefer those combinations that we like" (van der Waerden, 1953, p. 129; see also Poincaré, 1956, p. 2047–2048). A similar statement is also given by Weyl (Ebeling, Freund, & Schweitzer 1998, p. 209), who points out: "My work has always tried to unit the true with the beautiful and when I had to choose one or the other I usually chose the beautiful."

Thus theories, that have been described as extremely beautiful, as for example the general theory of relativity, have been compared to a work of art (Chandrasekhar, 1987); Feyerabend (1984) even considers science as being a certain form of art.

Mathematics and mathematical thought seem to be directed towards beauty as one profound characteristic. Papert and Poincaré even believe that aesthetics play the most central role in the process of mathematical thinking (see e. g. Dreyfus & Eisenberg 1986, p. 2; Hofstadter, 1979). Burton (2004, p. 88) points out that most of the mathematicians involved in a recent study rated highly the importance of aesthetics in their work. Nevertheless, this point of view is in general rarely considered. Davis and Hersh state (1981, p. 169):

> Blindness to the aesthetic element in mathematics is widespread and can account for a feeling that mathematics is dry as dust, as exciting as a telephone book, as remote as the laws of infangthief of fifteenth century Scotland. Contrariwise, appreciation of this element makes the subject live in a wonderful manner and burn as no other creation of the human mind seems to do.

Although aesthetics seem to play a crucial role in creative mathematical processes, there is a limited body of research on the meaning of aesthetics in the domain of mathematics. On the other hand, there is a substantial body of research on the nature of mathematical creativity.

Our question is what is the relationship between the two constructs? Before making this more precise (section 3), we go on with clarifying the concepts we focus on: general and mathematical creativity, and aesthetics.

AESTHETICS AND CREATIVITY: MEANING OF THE CONSTRUCTS

The concept of creativity is well-defined in literature. Creativity is a paradoxical construct to study because in many ways it is self defining. In other words, we are able to engage or judge acts of everyday creativity such as improvising on a recipe (Craft, 2002), use a tool in a way it wasn't intended, or intuit emotions and intended meanings from gestures and body language in day-to-day communication (Sriraman, 2009). Children are particularly adept at engaging in creative acts such as imaginary role playing or using toys and other objects in imaginative ways. "Aha!" experiences occur not only in individuals working on scientific problems but also in day-to-day problems such as realizing a person's name or relational identity after having forgotten it. However it is important to distinguish between everyday creativity and domain specific or paradigm shifting creativity. Domain specific creativity or "extraordinary creativity" causes paradigm shifts in a specific body of knowledge and it is generally accepted within that works of "extraordinary creativity" can be judged only by experts within a specific domain of knowledge. Some researchers have described creativity as a natural "survival" or "adaptive" response of humans in an ever-changing environment.

The Handbook of Creativity (Sternberg, 2000) which contains a comprehensive review of all research available in the field of creativity suggests that most the approaches used in the study of creativity can be subsumed under six categories: mystical, pragmatic, psychodynamic, psychometric, social-personality and cognitive. The mystical approach to studying creativity suggests that creativity is the result of divine inspiration, or is a spiritual process. In the history of mathematics, Blaise Pascal claimed that many of his mathematical insights came directly from God. This is somewhat analogous to the ancient Greeks belief in muses as a source of inspiration for artistic works. The pragmatic approach is focused on developing creativity. For instance, George Polya's emphasis on the use of a variety of heuristics for solving mathematical problems of varying complexity is an example of a pragmatic approach. The psychodynamic approach to studying creativity is based on the gestaltist idea that creativity arises from the tension between conscious reality and unconscious drives as popularized by Jacques Hadamard who constructed case studies of eminent creators such as Albert Einstein. The psychometric approach to studying creativity entails quantifying the notion of creativity with the aid of paper and pencil tasks such as the Torrance Tests of Creative Thinking developed by Paul Torrance. These tests are used by many gifted programs in middle and high schools, to identify students that are gifted/creative and show traits of divergent thinking. The test is scored for fluency, flexibility, and the statistical rarity of a response. Some researchers also call for use of more significant productions such as writing samples, drawings, etc to be subjectively evaluated by a panel of experts instead of simply relying on a numerical measure. The social-personality approach to studying creativity focuses on personality and motivational variables as well as the socio-cultural environment as sources of creativity. Finally the cognitive approach to the study of creativity focuses on understanding the mental processes that generate new and novel ideas. Most of the contemporary literature on creativity suggests that creativity is the result of confluence of factors from the six aforementioned categories. Two of the most commonly cited confluence approaches to the study of creativity are the "systems approach" of Mihaly Csikszentmihalyi; and "the case study as evolving systems approach" of Doris Wallace and Howard Gruber (see Csikszentmihalyi, 1996; Wallace & Gruber, 1992).

The systems approach takes into account the social and cultural dimensions of creativity, instead of simply viewing creativity as an individualistic psychological process and studies the interaction between the individual, domain and field. The field consists of people who have influence over a domain. For example, editors of research journals would have influence on any given domain. The domain is defined a cultural organism that preserves and transmits creative products to other individuals in the field. Thus creativity occurs when an individual makes a change in a given domain, and this change is transmitted through time. The personal background of individuals and their position in a domain naturally influence the likelihood of their contribution. It is no coincidence that in the history of science, there are significant contributions from clergymen such as Pascal, Copernicus and Mendel, to name a few, because they had the means and the leisure to "think." Csikszentmihalyi (1996) argues that novel ideas that result in significant changes are unlikely to be adopted unless they are sanctioned by a group of experts that decide what gets included in the domain. In contrast to Csikszentmihalyi's argument that calls for focus on communities in which creativity manifests, "the case study as evolving systems approach" treats each individual as a unique, evolving system of creativity and ideas, where each individual's creative work is studied on its own (Wallace & Gruber, 1992). The case study as an evolving system has the following components to it. First, it views creative work as multi-faceted. So, in constructing a case study of a creative work, one has to distill out the facets that are relevant and construct the case study based on the chosen facets. These facets are: uniqueness of the work, epitome (a narrative of what the creator achieved), systems of belief (an account of the creator's belief s system), modality (whether the work is a result of visual, auditory or kinesthetic processes), multiple time-scales (construct the time-scales involved in the production of the creative work), dynamic features of the work (documenting other problems that were worked on simultaneously by the creator), problem-solving, contextual frame (family, schooling, teachers influences), and values (the creator's value system).

Cultural and social aspects play a significant role in what the community, in general, and the school system, in particular, considers as "creativity" and how they deal with it. Numerous studies indicate that the behavioral traits of creative individuals very often go against the grain of acceptable behavior in the institutionalized school setting. For instance, negative behavioral traits such as indifference to class rules, display of boredom, cynicism or hyperactivity usually result in disciplinary measures as opposed to appropriate affective interventions. In the case of gifted students who 'conform' to the norm these students are often prone to hide their intellectual capacity for social reasons, and identify their academic talent as being a source of envy. History is peppered with numerous examples of creative individuals described as "deviants' by the status quo. Even at the secondary and tertiary levels there have been criticisms about the excessive amount of structure imposed on disciplines by academics as well as Euro-centric attitudes and male epistemology centered attitudes towards knowledge generation. Such a criticism particularly resonates in the world of science and mathematics, especially during elementary and secondary schooling experiences level, where minority, ethic minorities, first nation and female gifted/creative students are marginalized by practices that are alien to their own cultures (Sriraman, 2009).

Based on extensive classroom based research and informed by findings from the field of psychology and the history of science, five pedagogical principles to maximize general creativity in the classroom have been posited by Sriraman & Dahl (2009). The five principles are: (a) the Gestalt principle, (b) the Aesthetic principle, (c) the free market principle, (d) the scholarly principle, and (e) the uncertainty principle.

The Gestalt principle: Although psychologists have criticized the Gestalt model of creativity because it attributes a large "unknown" part of creativity to unconscious drives during incubation, numerous studies with scientists and mathematicians have consistently validated this model. In all these studies after one has worked on a problem for a considerable time (preparation) without making a breakthrough, one puts the problem aside and other interests occupy the mind. Jacques Hadamard put forth two hypotheses regarding the incubation phase: (a) The 'rest-hypothesis' holds that a fresh brain in a new state of mind makes illumination possible. (b) The 'forgetting-hypothesis' states that the incubation phase gets rid of false leads and makes it possible to approach the problem with an open mind. The Soviet psychologist Krutetskii explained that the experienced of sudden inspiration is the result of previous protracted thinking, of previously acquired experience, skills, and knowledge the person amassed earlier. This period of incubation eventually leads to an insight on the problem, to the "Eureka" or the "Aha!" moment of illumination. Most of us have experienced this magical moment. Yet the value of this archaic Gestalt construct is ignored in the classroom. This implies that it is important that teachers encourage the gifted to engage in suitably challenging problems over a protracted time period thereby creating the opportunities for the discovery of an insight and to experience the euphoria of the "Aha!" moment.

The Aesthetic principle: Many eminent creators have often reported the aesthetic appeal of creating a "beautiful" idea that ties together seemingly disparate ideas, combines ideas from different areas of knowlege or utilizes an atypical artistic technique. In mathematics, Georg Cantor's argument about the uncountability of the set of real numbers is an often quoted example of a brilliant and atypical counting technique

The Free market principle: Scientists in an academic setting take a huge risk when they announce a new theory or medical break through or proof to a long standing unsolved problem. The implication for the classroom is that teachers should encourage students to take risks. In particular they should encourage the gifted/creative students to pursue and present their solutions to contest or open problems at appropriate regional and state math student meetings, allowing them to gain experience at defending their ideas upon scrutiny from their peers.

The scholarly principle: One should embrace the idea of "creative deviance" as contributing to the body of knowledge, and they should be flexible and open to alternative student approaches to problems. In addition, they should nurture a classroom environment in which students are encouraged to debate and question the validity of both the teachers' as well as other students' approaches to problems. Gifted students should also be encouraged to generalize the problem and/or the solution as well as pose a class of analogous problems in other contexts. Allowing students problem posing opportunities and understanding of problem design helps them to differentiate good problems from poor, and solvable from non-solvable problems. In addition, independent thinking can be cultivated by offering students the opportunity to explore problem situations without any explicit instruction.

The Uncertainty Principle: Real world problems are full of uncertainty and ambiguity as indicated in our analysis so far. Creating, as opposed to learning, requires that students be exposed to the uncertainty as well as the difficulty of creating original ideas in science, mathematics, and other disciplines. This ability requires the teacher to provide affective support to students who experience frustration over being unable to solve a difficult problem. Students should periodically be exposed to ideas from the history of mathematics and science that evolved over centuries and took the efforts of generations of artists, scientists and mathematicians to finally solve. At the secondary school levels, one normally does not expect works of extraordinary creativity, however the literature indicates that it is certainly feasible for students to offer new insights into a existing/current scientific problems or a new interpretation or commentary to a literary, artistic or historical work.

While creativity, particularly mathematical creativity can be somewhat described in an objective way, feelings of aesthetics are very subjective. The subjectively experienced feelings of mathematical beauty are not so easily to be described. For this we have to bring out the characteristics of mathematical aesthetics. What does it mean, for example, that a theorem, a proof, a problem, a solution of a problem (the process leading up to a solution, as well as the finished solution), a geometric figure, or a geometric construction is beautiful?

Although assessments about beauty are very personal, there is a far-reaching agreement among scholars as to what arguments are beautiful (Dirac, 1977). Thus it makes a sense to search for factors contributing to aesthetic appeal. Before starting on this journey, Hofstadter (1979, p. 555) sounds a note of warning when suggesting, that it is impossible to define the aesthetics of a mathematical argument or structure in an inclusive or exclusive way, "There exists no set of rules which delineates what it is that makes a peace beautiful, nor could there ever exist such a set of rules."

However, we can find in the literature several indications of criteria determining the aesthetic rating (see also Brinkmann, 2000, 2004a, 2004b, 2006).

The Pythagoreans took the view that beauty grows out of the mathematical *structure*, found in the mathematical *relationships* that bring together what are initially quite independent parts in such a way to form a unitary whole (Heisenberg, 1985). Chandrasekhar (1979) names as aesthetic criteria for theories their display of "*a proper conformity of the parts to one another and to the whole*" while still showing "some *strangeness* in their proportion". Weyl (1952, p. 11) states that beauty is closely connected with *symmetry,* and Stewart (1998, p. 91) points out that *imperfect symmetry* is often even more beautiful than exact mathematical symmetry, as our mind loves surprise. Davis and Hersh (1981, p. 172) take the view that, "A sense of strong personal aesthetic delight derives from the phenomenon that can be termed *order out of chaos.*" And they add, "To some extent the whole object of mathematics is to create order where previously chaos seemed to reign, to extract structure and invariance from the midst of disarray and turmoil."

Beutelspacher (2003, p. 86) points out, that mathematical beauty is first *simplicity;* a mathematical description should be *brief* and *concise.* Whitcombe (1988) lists as aesthetic elements a number of vague concepts as: *structure, form, relations, visualisation, economy, simplicity, elegance, order.* Dreyfus and Eisenberg (1986) state, according to a study they carried out, that *simplicity, conciseness* and *clarity* of an argument are the principle factors that contribute to the aesthetic value of mathematical thought. Further relevant aspects they name are: *structure, power, cleverness* and *surprise.*

Cuoco, Goldenberg and Mark (1995, p. 183) take the view that the beauty of mathematics lies largely in the *interrelatedness of its ideas.* Ebeling, Freund and Schweitzer (1998, p. 230) point out, that the beautiful is as a rule connected with *complexity;* complexity is necessary, even though not sufficient, for aesthetics. In this context, the degree of complexity plays a crucial role, as e. g. a study with students carried out by Brinkmann (2004a) indicates: the permissible degree of complexity for a beautiful problem depends on the mathematical ability of each individual.

Complexity and simplicity are both named as principal factors for aesthetics: how do these notions fit together? If simplicity is named, it is mainly the simplicity of a solution of a complex problem, the simplicity of a proof to a theorem describing complex relationships, or the simplicity of representations of complex structures. It looks as if simplicity has to be combined in this way with complexity, in order to bring out aesthetic feelings (Brinkmann 2000).

We have to consider that all the quoted criteria for aesthetics are given by qualitative characteristics,[1] and hence by their nature they are fuzzy quantities. Thus aesthetic considerations will depend on individual judgements.

RESEARCH QUESTIONS

The focus of our interest was to explore the relationship of aesthetics and creativity in the field of mathematics. In other words, we first asked:

1. Do creative mathematicians necessarily also experience the feeling of mathematical beauty?
2. Is creativity a necessary condition for the aesthetic appeal?
3. Is an aesthetic appeal necessary for creative work?

As for question 3 it is clear, that the aesthetic appeal is not the only reason why researchers do mathematics.

Based on experience, we could definitively answer the second question by "no." Two examples may illustrate this.

Example 1: Learning mathematics in school is not necessarily done by creative processes. Students could, for example, just be informed about some mathematical contents and feel that they are beautiful. In a school lesson, e.g., there was shown a film on fractals to the students. There was no creativity demanded, the students just looked at the film and were fascinated!

Example 2: If mathematicians read an article in a mathematics journal they might experience some kind of mathematical beauty, without being involved in a creative process.

Thus we extended the second question by asking:

4. In which situations may mathematical beauty be experienced?

Further we asked, in more detail:

5. In which way success/failure in a creative problem solving process influences the former aesthetic appeal of the problem?

Inspired by (mostly historical) literature (see section 1), we wanted also to know:

6. If contemporary creative mathematicians deal the point of view, that beauty serves as an orientation in the work of mathematicians, and
7. If the aesthetic appeal for mathematics can be compared with that one for arts and music, and if arts inspired some creative endeavours in mathematics.

Additionally, we were interested:

8. If there is a hierarchy to the aesthetic appeal of mathematics and if so, what the relationship between this and the degree of corresponding creative work is.

METHODOLOGY

The aesthetic appeal of a mathematical object is not a characteristic possessed by this object but dependent on personal feelings. Thus we had to find out the relationship of aesthetics and creativity as it is personal experienced by contemporary mathematicians. For this, interviews with creative academics are a suitable means.

For reasons of practicability we decided to develop a questionnaire that should be sent and responded by e-mail. We addressed mathematicians in USA and Germany, known for notable works in different areas of mathematics. It was important for us to explore the statements of mathematicians working in the very different fields of mathematics in order not to get a one-sided view.

The questionnaire we used is shown in Figure 6.1. In Germany, mathematicians involved in the study received the questionnaire both in English and in German language and they wrote their answers in German. Thus semantic misunderstandings due to language problems were excluded to the extent possible.

The questions 1, 2, 4 and 5 in the questionnaire refer to question 7 in section 3. We expected that it might be difficult to classify aesthetics according to degree of appeal. The idea was, to use metaphors in the sense of comparing the aesthetical appeal of a mathematical object with that one of a nonmathematical object were the characteristic of beauty is more familiar ("This mathematical object is as beautiful as…"). Of course, the aesthetic rating of nonmathematical objects is also individually different. The first questions of the questionnaire is explained in order to explore individual levels of feelings of beauty (in every day life), and thus to get a sort of scale for aesthetic feelings. Question 2 in the questionnaire intends to integrate the aesthetic appeal of (certain) mathematics in the greater beauty hierarchy of the respective individual. Question 4 and 5 in the questionnaire examine a possible link between the degree of aesthetics and the degree of creativity; question 5 is based on the assumption that a researcher is more involved in creative working processes than a learner.

Question 3 in the questionnaire corresponds to question 2 in section 3; the questions 6 and 7 refer to the questions 1 and 3 in section 3. The research question 4 in section 3 is implemented in the questionnaire under 11, research question 5 under 8, 9 and 10, and research question 6 under 12.

The answers given by mathematicians should be analysed with regard to common or differing arguments.

RESULTS

The mathematicians involved in our study are working in very different fields of mathematics: algebra, discrete mathematics, geometry, non-classical and mathematical logic, history of mathematics. We have received the answers to our questions from eight mathematics researchers, these are about half

QUESTIONNAIRE

1. What do you feel is extremely beautiful/beautiful? Please give some examples. Can you order these examples according to their aesthetic appeal on you?

2. Did you experience some mathematical objects as beautiful? Which ones? Can you compare those feelings of mathematical beauty with the aesthetic appeal of the examples you gave above?

3. In which situations did you experience mathematical beauty? Were you involved in a creative process?

4. Do you see any link between the degree of creativity in your work and the degree of aesthetic appeal of the mathematics you dealt with? If so/If not please explain.

5. Can you compare/contrast your aesthetic appeal for the subject as you transitioned from a learner to a researcher of mathematics?

6. Can you describe a learning or a teaching moment when you experienced a sense of beauty for the subject matter? If so did it result in pursuit of research (in other words in a creative process)?

7. Was every creative work you did in mathematics connected with aesthetic feelings? Is(Was) an aesthetic appeal of the mathematics you deal(t) with necessary for your creative work?

8. Does/Did the aesthetic appeal of mathematics influence the choice of research questions?

9. Was your creative work lead by aesthetic feelings? Did aesthetic feelings play a role in your "choice"/route of finding a solution? When multiple solutions/solution paths were available did aesthetic feelings play a role for your choice of one solution over the others?

10. Dirac wrote about Schrödinger and himself: "It was a sort of act of faith with us that any questions which describe fundamental laws of nature must have great mathematical beauty in them." Can you agree with the similar statement: "It was a sort of act of faith in my work that any fundamental mathematical truth must have great beauty in it."

11. Does success/failure in a creative problem solving process influence the former aesthetic appeal of the problem? In other words, if the pursuit of a problem for an extended period of time does/did not result in an insight, does/did it diminish the aesthetic appeal in some way? On the other hand if the pursuit resulted in fruition... does/did it enhance or deepen the aesthetic appeal? Does success enhance the appeal of the problem that way, that it leads you to other related problems?

12. Can you compare your aesthetic appeal for the arts and music to your aesthetic appeal for mathematics? Have the arts inspired you in your creative endeavors in mathematics?

Figure 6.1 Questionnaire used in the study.

of the researchers to whom we asked to answer our questionnaire. With one exception, those who refused to answer our questions, argued this with time problems. One mathematician answered with no further comments: "that's not much good to me"; a comment hardly to be interpreted.

In seven of the eight responses we received, the aesthetical component is regarded as an important and influencing factor in the work of mathematicians; we will simultaneously present the given answers to our single research questions in detail.

However, one mathematician does not agree with a considerable function of aesthetics in the *daily work* of mathematicians. He states, that pure mathematics involves theories that are often very transparent and conclusive and therefore beautiful; but that in every-day work there is no place for questions about the aesthetic appeal of a working subject. He argues, that mathematicians cannot allow themselves the luxury of working on problems they like. In the work of a mathematician, creativity is generally demanded in problem solving processes for realistic applications. Contexts in which mathematics is honestly applicable to something, often involve "dirty" calculations and tiresome considerations. Thus, reality given, he sees other sources than the aesthetical appeal for mathematical creativity: e. g. curiosity or challenge by the problem. While working on a problem he tries to draft simple models, but the results often still involve ugliness. He does not see any link between the degree of creativity in the every-day work of mathematicians and the degree of aesthetic appeal of the mathematics they deal with.

We will now present the answers of the other seven mathematicians, named here M1 to M7.

First, some of them gave a sort of *definition* for their personal feelings *of mathematical beauty*, respectively some characteristics for mathematical aesthetics:

M1: I define beauty, roughly speaking, as the experience of realization (and surprise) that an extremely complicated phenomenon is actually just a special case of a general principal. This general principal might be a radically different perspective. This sort of experience occurs all the time in mathematics and physics.

M2: I feel what is beautiful is that which gives pleasure to the mind, the sight, the hearing or the heart, or a combination of these.

M4: I feel the simple and unexpected as particularly beautiful. More precisely, how you can get far with only very few assumptions. Usually, beauty is to equate with simplicity in my research work.

M6: A concrete mathematical model can hold an aesthetic appeal, as well as a virtual one. A particularly clear and concise new proof or just a particularly smart example can cause this feeling. Also the application of elegant mathematical methods for a practical problem can be joined by the feeling of beauty. Certainly, a particularly intensive appeal comes from the suddenly (and sometimes unexpected) pure discovery, the clear understanding of a mathematical phenomenon, often from a completely new perspective, a new harmonic interplay of different fields, first appearing not to be related to each other.

M7: Beauty within mathematics manifests itself on the one hand by typical mathematical-logical arguments, especially if these arguments show unexpected and important connections, in an (at first) surprising manner and then mostly also in a surprising simple manner. On the other hand, mathematical beauty shows itself in the structures used by mathematicians, for me especially in algebraic structures. To be precise, first especially then, if these structures fit extremely good to an important problem class and are at the same time so abstract that the treatment of the relevant (non-trivial) problems becomes "simple". Second, within the treatments of such structures them themselves, especially then, if the conceptual grasp of these structures is as advanced, that an "elegant" solution of structure internal problems is possible.

As for the answer to *question 1* there were listed non-mathematical examples for things or situations felt as extremely beautiful, e.g.:

M2: Seeing Bryce Canyon and hiking it in the cool of the day; watching a swan on a lake; observing a cat at play, or play with it; listening to a symphony or playing a musical instrument well; absorbing words fitly spoken; experiencing a well-composed and delivered lecture or sermon; tracing the lines of a child's face or the lines in the hands of an elderly person; experiencing fireworks on the 4th of July; seeing the flag of the USA flying in the breeze; watching a sunset or a jet stream float across the sky,

M3: Prince William Sound, Quadratic Reciprocity, Bach's Toccata and Fugue in D minor, Hamlet, Golden Retrievers,

M5: Nearly everything in the field of e-music: Gregorianik, Monteverdi, Bach, Mozart, Beethoven, Schubert, Schumann, Brahms; literature: Fontane (everything), H. M. Enzensberger, a lot of poems; painting; beautiful, quiet landscapes; beautiful, intelligent women;

M7: Well played classical music; close to nature mountainous landscapes; a warm summer evening on the terrace of the own house, together with my wife and a good bottle of wine; intelligent conversations with good friends.

But, no one ordered his examples; a ranking seems to be very difficult or quite impossible in this field. This applies accordingly for the answers given to *question 2*. Thus, we did not succeed with our initial idea to classify aesthetics according to degree of appeal by using metaphors in the sense of comparing the aesthetical appeal of a mathematical object with that one of a nonmathematical object (see section 4). Nevertheless we experienced examples to what affects mathematicians, both, in non-mathematical areas (see above) and in the field of mathematics, as follows:

M1: Prime examples are Newton's theory of gravitation usurping the theory of epicycles. An example in mathematics is the notion that number theory and the classical theory of algebraic geometry are two aspects of the theory of schemes.

I would not say that I experienced actual objects as beautiful. Instead, I find that certain mathematical objects naturally facilitate a new, more general perspective, which allows one to answer old questions in a new way... such examples abound in mathematics.

One example is the Lebesque theory of integration. The theory provides a different perspective of integration than the Riemann theory, and makes certain questions related to the compatibility of the integral and the sum much easier to deal with.

Another example is the application of the theory of cohomology to the theory of schemes. Using a complicated construction called Etale cohomology, Grothendieck's school was able to answer questions of Weil regarding objects whose definitions seemed to have nothing to do with cohomology.

In both examples... there is an experience of great surprise that such a novel new theory can be used to answer old questions whose statements have nothing to do with the new theory.

M2: Reading or writing or explaining an elegant proof of a theorem, or an elementary proof of a great theorem, or proofs which involve the interaction or algebra and geometry or other branches of mathematics coming together, or...

M3: Quadratic Reciprocity, Godel's theorem, Fundamental theorem of algebra, Church-Turing thesis, Riemann zeta function.

M4: Robinsons Q; the proof of the prime number theorem by Cornaros and Dimitracopulos within a week number theory; Erdoes' proof of the Tschebyscheff's theorem, that there always exists a prime number between n and 2n, and Paolo D'Aquino's proof, that this may be implemented in a week number theory; my definition for two perpendicular circles, using only "there exists," "and," "unequal," and the "point-circle incidence"; the whole numbers and their simplest properties, about which we hardly know something—they are probably the most beautiful objects of mathematics; in addition, simple statements, simple axioms, unexpected relations.

M5: Within mathematics I feel relations between objects as beautiful. In the field of mathematics a particular relation often proves to be something that "fits." We feel this as being beautiful. This reaches from simple examples as: "the equation is solvable" over particular ways of solution, but also to particularly comprehensive structures (pentagon, soccer ball . . .).

M6: Actually, mathematics as a whole seems to me to be characterized by particular aesthetic, and one could give a nearly unlimited list of mathematical phenomena of great beauty.

One mathematician (M5) pointed out, that on the one hand one can compare the aesthetic appeal of mathematical objects with that one of non-mathematical things, but on the other hand, within mathematics, there appears in the moment of the awareness of the fitting also a particular verity, that cannot be taken away again. This, he does not experience as much in arts.

Another mathematician (M6) stated that "the magnificent sight from a mountain peak on the mountain scenery of the Alps is, after a strenuous hiking tour, even more overwhelming" than mathematics.

When asked if there is seen a link between the degree of creativity in one selves work and the degree of aesthetic appeal of the mathematics dealt with (*question 4*), most of the interviewed mathematicians answered that they did so:

M1: I do see such a link. My feeling is that, the more esoteric and general the new perspective, the more beautiful it is. If I (or someone else) must develop extremely abstract (and seemingly irrelevant) machinery in order to answer classical questions which seem to have nothing to do with that machinery, than it surely takes a very new perspective to see the link between the old and the new. Thus, the paradigm shift is much more extreme, and the experience of realizing that the old phenomenon is a very special case of the new theory is that much more surprising.

M4: Yes, there exists a very significant link. If I would not have found an aesthetically fascinating part of mathematics, I would have probably stopped a long time ago to do research, or even would have left mathematics totally. Only to prove something because others are interested in it and maybe one could make a career for oneself—and could keep one's head above water—is not enough motivation for me, to wake me out of my natural laziness.

M5: In my research work I have been, nearly always, only interested in those new findings, which were aesthetically satisfying. Findings that are "only true," with which we do not "feel" why they are true, are boring.

M6: The intensity of the emotional experience when suddenly all pieces of the puzzle of a new mathematical discovery fit together, is surely related to the intensity of the creative study of the respective topic.

Nevertheless, we have got also different answers. M2 responded question 4 by "no" and M7 replied: "I never thought about this: afterwards, an answer, on the truth of which you could rely, is not possible for me."

The comparison of the aesthetic appeal for the subject as a learner with that one as a (more creative working) researcher, demanded by *question 5*, emphasizes in some cases the answers given to question 4:

M1: As a learner of the subject, one experiences new perspectives on a weekly basis, so one sees the general "explain" the more specific relatively often. However, as a researcher, one develops a much deeper faith in the fact that a novel perspective can have enormous power.... I had to have a great deal of faith that my more general theory would work, since I spent many months developing it before knowing whether it would answer the question I wanted it to answer. I am still amazed that the theory answers the question I set out to answer, and this deepens my faith in the power of beauty as I have defined it.

M5: The experience to rediscover a connection that was already drawn by a great scientist is an almost equivalent experience [with respect to aesthetic feelings]. It is "weaker," as it is not one's own, but as a rule it is much more "stronger," as it concerns something much more important or may concern much deeper relations.

M6: The feeling of beauty . . . develops further while working intensively.

The answers given to the *questions 3, 6, and 7* show situations in which mathematical beauty has been experienced during a creative process. But, they show also that creative processes are neither necessary nor sufficient for the experience of mathematical beauty: on the one hand, there exist creative processes, which are not accompanied by aesthetic feelings, on the other hand aesthetic feelings may arise in situations in which one is not involved in a creative process (as already marked in section 3). The following extract of the given answers give more insight:

M1: I have experienced mathematical beauty many times, in particular, whenever I learn new mathematical theories that answer old questions, or answer questions unexpectedly. I have also experienced mathematical beauty in the process of solving problems. For example, in trying to solve a problem about a type of algebraic surface, I attempted again and again to use old tools and techniques. The old ways failed me. In desperation, I started reinventing some of these old tools in a more general context. To my amazement (after several months of developing this new theory), my invention turned out to be exactly the right tool to answer the questions I wanted to answer. This is an example of a new perspective making old questions answerable. This sort of thing does not always happen when tackling a mathematics problem. Much of the time, all that is required is the clever and agile use of old tools.

M2: I have experienced mathematical beauty in reading, in hearing lectures, in talking with students/colleagues, in working on research. . . . Often I was not involved in a creative process.

M3: [I have experienced mathematical beauty in] studying, teaching, doing research.

It has been a cumulative process—no particular moment. Both teaching and research have contributed to my appreciation of the beauty of mathematics.

M4: Not every creative work was connected with aesthetic feelings. Most of these works were accompanied by nail-biting and unpleasant feelings, sleeplessness and things like that. Only after having proved the theorem, one could speak about the aesthetic effect of the result.

M6: Even if a difficult proof for a statement is finally produced, one is not satisfied until its inherent aesthetic really reveals the context to be expressed. Also in mathematics creativity does not necessarily lead to aesthetics, but the aesthetic claim is a great motivation for one's creative work.

In *question 7* we have asked also if an aesthetic appeal of the mathematics one deals or dealt with is/was necessary for creative work. As to this point, the answers differed—some were agreements, others negations:

M1: Yes, since the experience of creativity is tied to the experience of changing ones perspective. Aesthetic appeal is necessary to me for my creative work since the process of changing perspective is quite painful to me. I have learned, however, that the payoff (i. e. the aesthetic appeal I experience) is more than ample payment for the labor of altering my perspective.

M2: I would not say aesthetic appeal was necessary for my creative work.

M3: No, some research was driven by practical considerations, e.g. my PhD thesis, tenure, but later work by intellectual interest which I equate with aesthetic feelings. The work of which I am most proud was driven by intellectual interest.

M4: My entire research work aims exclusively the hidden beauty of geometrical theories.

M6: There exists a plenty of such experiences [of mathematical beauty]; they are the researcher's actual drive.

M7: No.

The answers of the interviewed mathematicians to *question 8*, referring to aesthetic appeal as a possibly guiding principle when choosing a research question, all pointed in the same direction, whereas *question 9*, referring to aesthetics as possibly having a leading function when choosing a route of finding a solution, or one solution over other possible ones, provided diverse answers.

The mathematicians affirmed the influence of the aesthetic appeal on their choice of research questions, some everywhere, some partly. M6 for example stated: "The already recognizable or through new findings expected aesthetic is a great drive for the choice of the research subject." Two mathematicians named some restrictions:

M1: It does, but not completely. The reason is that even modest questions which do not seem to have a great deal of aesthetic appeal lead to beautiful and unexpected insights. This has happened time and again in the history of mathematics. I don't find any intrinsic appeal in the statement of Fermat's last theorem. However, the enormous depth of the mathematics it has inspired is dense with beauty.

M2: I would say more of my work is motivated by just wanting to know the truth.

Regarding the question if creative work (the choice of solutions / solution paths) was lead by aesthetic feelings, the following answers were given:

M1: It has been lead by aesthetic feelings, and in my choice of finding a solution, much to my detriment. Let me explain: I have had some success in picking aesthetic approaches to solving problems. Furthermore, in one important case. . . . aesthetics was my only guide to finding a solution. The idea for the solution was very simple, but my faith had to be very great indeed in order to build a theory around this idea to make sure it actually worked. I had never before experienced such a powerful affirmation of faith. During a more recent project, my experience led me to believe that beauty was my primary guide for discovering mathematical truths. This led me astray for many months.

M2: No. Not especially.

M3: No, selecting technique was more a practical matter of that which with I was familiar or able to study and learn. Ideas are tried and discarded mostly based on how well they seem to fit.

M4: Actually, no. Finding a solution for any, however aesthetic result is matter of trench war. One stands deep in the mud and does not come out until the supposition is proved.

M5: Clearly. Aesthetics means in mathematics at first „simplicity". We try to get complex situations under control by trying to strive for the point from that everything becomes easy. Insofar every substantial mathematical work has to assume this aesthetic principle.

M6: If alternatives for a proof are visible, one would generally decide for the way that appears to be more elegant.

M7: Of course: if you see several ways of solution for a problem, you take that one, which appears as more elegant.

Four mathematicians agreed with the statement "It was a sort of act of faith in my work that any fundamental mathematical truth must have great beauty in it." (*question 10*), while three mathematicians negated it:

M1: By my definition of beauty, this is true. Since I have defined beauty to be the realization that a new (general) perspective explains older mathematics, and since one can replace "fundamental" by "very general", the statement in the question is "It was a sort of act of faith in my work that any very general mathematical perspective must have great beauty in it." This is almost a tautology.

M2: Not especially. I would not regard DeBranges' proof of the Bieberbach conjecture as particularly beautiful, but it IS true.

M3: No. There remain fundamental results—e.g. the four color solution; the solution of Fermat's conjecture—that are fundamental but (at this time) lack beauty.

M4: Yes.

M5: Absolutely. Hardy said (cited from memory): Beauty is the ultimate goal. . . . There is no permanent place for ugly mathematics.

M6: If someone feels the clarity of structures as aesthetic value, the search for underlying structures ("life, universe and everything") is not separable from the search for inherent aesthetic.

M7: No.

By *question 11* we wanted to know, if failure respectively success in a creative problem solving process influences the former aesthetic appeal of the problem. Here we list the very different answers:

M1: Fruition often leads to new insights which put the original question in a larger context. This can lead to generalizations, which are inherently more beautiful.

Certainly, success enhances ones understanding of the problem, and hence naturally suggests related, tractable problems. And tractability plays an important role in what mathematics gets done.

M2: I don't think success or failure changes the aesthetic appeal.
M3: It seems to me, failure makes the problem more attractive—e.g., Riemann hypothesis, twin prime conjecture, 3n + 1 problem, P/NP problem. (I've worked on the latter two.)
M4: Of course, failure influences the direction of future projects, but this is not the case for the aesthetic estimation of that what had to be proved but could not be proved.

Success always leads to a deepening of the subject and to similar questions, as one hopes to stand a better chance with such questions.

M5: This question does not hit the mark. The feeling of beauty arises with the discovery of a new, simple, unexpected…connection. Insofar: Only a solution brings the flow.
M6: Of course, the disappointment about the fact that a supposed intrinsic aesthetic did not reveal, in spite of intensive efforts, has an effect, and generates doubts, if the expected special beauty of the context really exists. The felt aesthetic of a scientific finding reinforces positively the own activities.
M7: Problems never have had an aesthetic appeal to me: one meets problems or is confronted with problems. If they do not arise directly out of practical interests, they may appear more or less "natural", but they withdraw (for me) from classification according aesthetical points of view.

Thus, we can state no changes of aesthetic appeal in some cases, but also enhanced attractiveness as a result of success, and even enhanced attractiveness as a result of failure in other cases. However, to respond question 11 turned out to be problematic for two mathematicians, because of their personal definition/feeling of aesthetics.

Regarding *question 12*, the mathematicians M1 to M5 gave very similar statements. They pointed out analogies between their aesthetic appeal for the arts and music and that one for mathematics, but stated—like M6 and M7 too—that the arts have not inspired them in their creative endeavours in mathematics. Here are some excerpts of the answers:

M1: In thinking about mathematics, one often loses self-consciousness and forgets about the everyday minutiae of life. The state of concentration which is often required for learning about or discovering mathematics is akin to a meditative state: ones ordinary senses are ignored. This is a similarity between the aesthetic appeal for the arts/music and for mathematics.

On the other hand, mathematics is far less forgiving than the arts. Just about any "out of the ordinary" feeling inspired by the arts or music is popularly tied with the aesthetic of the subject. To appreciate mathematics takes a very particular and intense effort, as well as a great deal of training.

For this reason, the arts have never inspired me in my creative endeavors in mathematics.

M2: Yes, in so far as I think of form and structure. Some art and music is chaotic—mathematics usually tries to describe chaos, not just sense it or absorb it. The arts have not inspired me in my creative work in mathematics.

M5: (a) Yes. In both cases one gets access to worlds, that otherwise would not be accessible. (b) No, I have not been inspired by music or the like in a more or less direct way.

M6: As my examination of music and arts is less intensive than that of mathematics, the emotionality of my feelings for inspiring works in music or arts follows another curve as for mathematical inspiration.

M7: These are separate worlds for me.

CONCLUDING REMARKS

The results of our study show, that the aesthetic appeal seems to play a crucial role in the creative work of contemporary mathematicians. As Nathalie Sinclair (see Sinclair, 2001, 2004, 2006 a, b, 2008) has pointed out, the question of "aesthetics" has been regarded as either frivolous or elitist in both mathematics and mathematics education. The elitist perception of the aesthetic dimension in mathematics education is attributable to the fact that most descriptions of "mathematical beauty" found in the literature come from eminent mathematicians. However classroom based research has shown that students are capable of appreciating the beauty inherent in mathematics (see Brinkmann, 2000, 2004a,b, 2006). In our study, it was found that the aesthetic component need not necessarily derive or be connected to a theorem or proof that the mathematician is currently working on, which can more often be one of sustained trial and frustration, described by one of the mathematicians as "trench war", but aesthetics is often present in appreciation of other results, reading elegantly presented material in books as well as listening to lectures from peers.

Aesthetics has also been relegated by some mathematics education researchers as a small component of the affective dimension of learning, when in fact it intertwines with both the cognitive and the affective components, and as indicated by our study an important aspect of creativity. Sinclair (2009) also presents a convincing argument that aesthetics may very well be the missing gap in numerous failed attempts at motivating students. Given the numerous reform movements in mathematics education that have occurred in many parts of the world, and the call to view school students as budding mathematicians and to get them engaged in mathematical thinking, it is ironic that the aesthetics has not received more attention by the community of mathematics educators. The five pedagogical principles outlined earlier give one possible way in which creativity can be fostered in the classroom. John Mason in his numerous writings has often offered insights into "*the lived experience of mathematical thinking*", and we think we are justified in claiming that aesthetics forms an important aspect of this lived experience!

NOTE

1. Birkhoff (1956) made an attempt to quantifying aesthetics in a general way, but his proposal seems not to be very convincing.

REFERENCES

Beutelspacher, A. (2003). *Mathematik für die Westentasche*. München und Zürich: Pieper.

Birkhoff, G. D. (1956). Mathematics of aesthetics. In J. R. Newman (Ed.), *The world of mathematics* (Vol. 4, 7th ed., pp. 2185–2197). New York: Simon & Schuster.

Brinkmann, A. (2000). Aesthetics—complexity—pragmatic information. In S. Götz & G. Törner (Eds.), *Research on mathematical beliefs. Proceedings of the MAVI-9 European Workshop, June 1–5, 2000, in Vienna* (pp. 18–23). Duisburg: Schriftenreihe des Fachbereichs Mathematik SM-DU-482.

Brinkmann, A. (2004a). *The experience of mathematical beauty*. ICME-10, The 10th International Congress on Mathematical Education, July 4–11, 2004, in Copenhagen, Denmark. http://www.icme-organisers.dk/tsg24

Brinkmann, A. (2004b). Mathematische ásthetik—Funktionen und Charakteristika des Schönen in der Mathematik. In A. Heinze & S. Kuntze (Eds.). *Beiträge zum Mathematikunterricht 2004*, (pp. 117–120). Hildesheim, Berlin: Franzbecker.

Brinkmann, A. (2006). Erfahrung mathematischer Schönheit. In A. Büchter, H. Humenberger, S. Hußmann, & S. Prediger (Eds.). *Realitätsnaher mathematikunterricht—von Fach aus und für die Praxis. Festschrift für Hans-Wolfgang Henn zum 60. Geburtstag*, (pp. 203–213). Hildesheim, Berlin: Franzbecker.

Burton, L. (2004). *Mathematicians as enquirers learning about learning mathematics*. Boston, Dordrecht, New York, London: Kluwer Academic.

Chandrasekhar, S. (1973). Copernicus—From humanism to inquisition. *Bulletin of the Atomic Scientists 29*, 6, 27–30.

Chandrasekhar, S. (1979). Beauty and the quest for beauty in science. *Physics Today 32*, 25–30.

Chandrasekhar, S. (1987). The aesthetic base of the general theory of relativity. In S. Chandrasekhar (Ed.), *Truth and beauty. Aesthetics and motivations in science*, (pp. 144–170). University of Chicago Press.

Craft, A. (2002). *Creativity in the early years: a lifewide foundation*. London: Continuum.

Csikszentmihalyi, M. (1996). *Creativity*. New York: Harper.

Cuoco, A. A., Goldenberg, E. P., & Mark, J. (1995). Connecting geometry with the rest of mathematics. In P. A. House, & A. F. Coxford (Eds.), *Connecting mathematics across the curriculum. 1995 yearbook of the national council of teachers of mathematics*, (pp. 183–197). Reston, VA: The Council.

Davis, P. J., & Hersh, R. (1981). *The mathematial experience*. Boston: Birkhäuser.

Dirac, P. (1977). *History of twentieth century physics*, (p. 136). Proceedings of the international School of Physics "Enrico Fermi", Course 57, New York: Academic Press.

Dreyfus, T., & Eisenberg, T. (1986). On the aesthetics of mathematical thought. *For the Learning of Mathematics—An International Journal of Mathematics Education, 6*(1), 2–10.

Ebeling, W., Freund, J., & Schweitzer, F. (1998). *Komplexe strukturen: Entropie und information*. Stuttgart, Leipzig: Teubner.

Feyerabend, P. (1984). *Wissenschaft als Kunst*. Frankfurt a. M.: Suhrkamp.

Ghiselin, B. (1952). *The creative process*. NY:The New American Library.

Heisenberg, W. (1985). Die bedeutung des schönen in der exakten naturwissenschaft. In W. Heisenberg (Ed.), *Gesammelte werke*, (pp. 369–384). Band C III, München: Piper.

Hofstadter, D. R. (1979). *Gödel, Escher, Bach: An eternal golden braid*. New York: Basic Books.

Poincaré, H (1956). Mathematical creation. In J. R. Newman (Ed.), *The world of mathematics*, (Vol. 4, 7th ed., pp. 2041–2050). New York: Simon & Schuster.

Sinclair, N. (2001). The aesthetic *is* relevant. *Learning of Mathematics 21*(2), 25–32.

Sinclair, N. (2004). The roles of the aesthetic in mathematical inquiry, *Mathematical Thinking and Learning, 6*(3), 261–284.

Sinclair, N. (2006a). *Mathematics and beauty: Aesthetic approaches to teaching children.* New York: Teachers College Press.

Sinclair, N. (2006b). The aesthetic sensibilities of mathematicians. In N. Sinclair, D. Pimm, & W. Higginson (Eds.), *Mathematics and the aesthetic: New approaches to an ancient affinity* (pp. 87–104). New York: Springer.

Sinclair, N. (2008). Attending to the aesthetic in the mathematics classroom. *Learning of Mathematics, 28*(1), 29–35.

Sinclair, N. (2009). Aesthetics as a liberating force in mathematics education? In press, *ZDM-The International Journal on Mathematics Education, 41*(nos 1&2).

Sriraman, B. (2009). General creativity. In press in B. Kerr (Ed). *Encyclopaedia of giftedness, creativity and talent.* CA: Sage.

Sriraman, B., & Dahl, B. (2009). On bringing interdisciplinary ideas to gifted education. In press in L.V. Shavinina (Ed). (pp. 1235–1256). *The International Handbook of Giftedness.* Springer Science.

Sternberg, R. J. (Ed.) (2000). *Handbook of creativity.* Cambridge University Press.

Stewart, I (1998). *Die Zahlen der natur: Mathematik als fenster zur welt.* Heidelberg, Berlin: Spektrum, Akademischer Verlag.

van der Waerden, B. L. (1953). Einfall und Überlegung in der mathematik. *Elemente der Mathematik VIII*(6), 121–144.

Wallace, D. B., & Gruber, H. E. (1992). *Creative people at work: Twelve cognitive case studies.* Oxford University Press

Weyl, H. (1952). *Symmetrie.* Basel, Stuttgart: Birkhäuser.

Whitcombe, A. (1988). Creativity, imagination, beauty. *Mathematics in School 17*(2), 13–15.

CHAPTER 7

THE MATHEMATICAL STATE OF THE WORLD

Explorations into the Characteristics of Mathematical Descriptions

Ole Ravn Christensen, Ole Skovsmose
Aalborg University, Denmark

Keiko Yasukawa
University of Technology, Sydney

ABSTRACT

In this chapter we try to analyse the conditions for describing the world mathematically. We consider the role played by mathematics in discussing and analysing "the state of the world." We use this discussion to clarify what it means to use a mathematical description. We illustrate why the concepts of "mathematical description" and "mathematical model" are inadequate to evaluate the use of mathematics in decision-making processes. As a result we develop a conceptual framework that is complex enough to match what goes on in scenarios involving applications of mathematics.

Relatively and Philosophically E^arnest, pages 81–94

81

Mathematics is a powerful tool. It influences our political decision-making in both process and outcome. A mathematical description of a given situation could, for instance, help us decide whether we should build more kindergartens, whether we should build a bridge in the cheaper way or the more expensive way, how many elderly people per hour an aged care worker should be expected to assist to take a shower, and so on. Therefore, the quality of the mathematical description that is involved in decision making becomes important.

But how do we define "quality" in relation to a mathematical description? One could argue that a good mathematical description is one that is based on a theoretically correct analysis of numerical data; that is, the quality depends on the mathematical treatment of numbers being rigorous, consistent and accurate. Many decisions of political interest and significance are motivated, however, by more than simple mathematical relations between data; for example many economic decisions refer to complex economic models, and environmental policies are founded on models that incorporate interpretations of what an ecological balance of the environment could mean.

This leaves us with the question of whether the concept of quality, such as what we have alluded to above, in relation to mathematically based decision making in today's political environment is an adequate one. It is a very common belief that the application of mathematical descriptions in decision making is pretty much a straight forward matter because mathematics by way of its nature deals with the essential structures of the world's phenomena. When debating the quality of a mathematically based decision, only questions concerning the rigour of the mathematical description and analysis appear to be open to serious discussions.

But the problem of quality regarding mathematical models and descriptions may need to be considered in other lines of reasoning all together. We could even imagine that there could exist phenomena which a mathematical description would never be able to account for, even when it is elaborated in the most detailed way and complies with all possible demands of mathematical exactitude. For example, environmental issues can generate a wide range of reactions from groups and individuals of different political inclinations. Some of their responses are expressed in technical and scientific terms; some in purely economic terms; still others in socio-cultural terms. If political decisions privilege mathematically rigorous arguments, then those arguments which are not "mathematisable" because of their nature necessarily get dismissed. For example, there may be a sacred site for an indigenous population that is known also to be rich in minerals. It may well be possible to analyse the economic costs and benefits of mining that site through a detailed mathematical description; however, it is both inappropriate and impossible to "mathematise" the cultural significance of the site.

Therefore, one could think of the problem of the quality of a mathematical description in the following way: on the one hand, one could imagine that a mathematical description of some aspect of the real world could depict essential elements of what is being described, and that mathematics in this way could provide a deep insight in the basic structures of a situation which otherwise would not have been identified. In this way mathematics could help to provide a basis for decision making. On the other hand, one could imagine that a mathematical description would be limited and impose a particular perspective on what one is seeing. Thus, a mathematical description turns into a prefabricated construct of what one is seeing. The consequence could be that mathematics-based decisions reflect, not just a particular type of deep insight, but also a certain rationality which is expressed by the mathematical formalism.

In this chapter we try and analyse the foundation for describing the world mathematically. We set up the task to clarify what it means to describe with mathematics; we aim to illustrate why the concept of a "mathematical description" or "mathematical model" is inadequate to evaluate what goes on in the application of mathematics in decision making processes; and finally we seek to develop a conceptual framework for the application of mathematics that is complex enough to match what goes on in general application scenarios involving mathematics.

PRIMARY AND SECONDARY SENSE QUALITIES

During the Renaissance the idea that the phenomena of the world have two distinctly different types of qualities was emphasised. Thus, Galileo Galilei differentiated between primary and secondary sense qualities. We experience the secondary qualities as taste, colour, sound, etc. These qualities depend on the person who perceives the object: the way one tastes some food refers to personal *experiences* of a given thing. To another person these experiences could be rather different. In fact, it can be difficult to define in what way one can even compare experiences of secondary qualities. The implication is that one should be sceptical of any insights about the natural world that is founded on the secondary qualities because there can be no agreed reference points. The primary qualities, on the other hand, refer to properties that can be measured. These qualities include weight, height, volume, position, movement, speed etc. Now the insight of Galileo and many other scientists in early modern science was that the primary qualities were objective qualities—they could be measured and everybody would agree about the measured results. The primary qualities of objects represent the "objective" properties of the world. In contrast to this the secondary are subjective qualities that depend on the perceiving subject.

In many cases the primary qualities, observable through mathematics, are, apart from being "objective" also "hidden" qualities that are unobservable within an everyday personal experience. For instance, it would be impossible for us to argue how many people smoke or how dangerous smoking really is from just wandering around in our everyday environment and not paying attention to the primary qualities of things. Many issues can only be resolved through a mathematical treatment that expresses results in numbers. With the realisation of the primary qualities of things Modern Science discovered a realm of *hidden truths* about the world that could be described without reference to subjective experiences. The secondary qualities of things, on the other hand, were elusive and hard to pin down in a way that could reveal objective, incontestable truths.

This division between primary and secondary qualities therefore carved out a clear territory for science to explore, namely the primary qualities of things. The primary qualities could be measured and thereby expressed through mathematics. This means that mathematics was afforded a particular role in the formulation of insights about the natural world. In Galileo's view, mathematics played a particular role in this formulation. Mathematics became the language of modern science. Mathematical descriptions of the world were in themselves valuable because they dug out yet undiscovered truths about the world that we could not perceive through our everyday experiences. He says:

> Philosophy is written in this grand book, the universe, which stands continually open to our gaze, but the book cannot be understood unless one first learns to comprehend the language and read the letters in which it is composed. It is written in the language of mathematics, and its characters are triangles, circles, and other geometric figures without which it is humanly impossible to understand a single word of it; without these, one wanders about in a dark labyrinth. (Galileo cited in Crosby, 1997, p. 240)

The related idea that mathematics was a joint language of all sciences has especially been pursued since the end of the 19th century. Gottfried W. Leibniz, however, at an early stage mentions that science should strive for formulating its knowledge through a unified language of logic. Eventually such a language was developed by Gottlob Frege and Bertrand Russell who tried to show that mathematics is exactly this unified language of logic that could be used by all sciences to describe the primary qualities of things. This idea was celebrated and further developed by logical positivism, not least through the work of Rudolf Carnap, who paid much attention to the nature of the mathematical language through which scientific insight should be formulated.

Modern Science has been extremely successful. A strong component of this success is the move away from natural philosophy trying to express the essential qualities of nature by ways of qualitative studies to focusing on the measurable mathematical relations between the phenomena of the world. But in addition to this move Modern Science has relied on the assumption that all knowledge can, and ideally should be, mathematised. If you have a problem to solve you had better start taking measurements because until you have done so you have not really treated it scientifically. This conception of knowledge and science has not only permeated the natural sciences but also the humanities and especially the social sciences. Thus, Emile Durkheim was highly inspired by Auguste Comte's positivism, while people like Otto Neurath, Ernst Nagel, Talcott Parsons and many others argued for establishing the social sciences according to the scientific paradigm exercised in natural science: Science is measuring. Science is the ability to put your problem into a mathematical model.[1]

In what follows we will discuss the special role attributed to mathematics as a pillar of Modern Science. The discussion will be motivated by an important research question, namely the question: *what is the state of our planet?* Is the state of the world gradually improving or are we in fact experiencing a world that is gradually becoming more and more uninhabitable? And what can a mathematical model tell us about this question? And further, what can this example tell us about the role of mathematics more generally? As mentioned earlier, mathematics can be considered to be the primary language of Modern Science, but what does it mean to formulate problems, ideas and solutions in this language? So we can ask what the mathematical discourse does to the way we *see* the world, and the way we *act* in the world.

WHAT IS THE STATE OF THE WORLD?

It appears that "the real state of the world" provides a necessary foundation for all kinds of overall decision making concerning global environmental issues. However, what is the state of the world, and how do we perceive this state? One analysis of the state of the world is provided annually by the World Watch Institute through its *State of the World* publications.[2] These and many other similar publications of whatever political leanings rely heavily on statistical data to argue their case—which sometimes is for radical change in the way we live, sometimes for "do nothing," and sometimes for something in between. We will take a particular example of a state of the world publication to raise some concerns about using a mathematical analysis of the state of the world in an unreflective way.

In 1998, Bjørn Lomborg provoked considerable controversy with his book, *Verdens sande tilstand*, and it was later followed by a revised edition in English, *The Skeptical Environmentalist—Measuring the Real State of the World* (2001), that triggered international attention. The theme of the book is the global environmental debate, which is seen by Lomborg as being dominated by what he calls the "litany" of doom and gloom. The picture that is spread through the news media on the state of the world is, according to Lomborg, one without good news. We are presented with catastrophes of hunger, hurricanes, stories about cases of devastating pollution among many other horror scenarios because of, in Lomborg's view, a basically unfounded belief that the world is going to hell! Lomborg is keen to convey to the public that there is no basis for the view that there is a global environmental crisis, and he wants to convey that on the contrary, the environment is actually getting better all the time.

Lomborg analyses a number of subjects, e.g. hunger, pollution, extinction of species and waste management problems, and applies statistical methods to corroborate his claim that things are going better than what one would believe from the litany of doom. It is an important aspect of Lomborg's work that it is not new numbers that he is working with but the same numbers as those his opponents in the environmental debate have worked with, those data sourced from big international organisations such as the Food and Agriculture Organisation and the World Health Organisation of the United Nations (Lomborg, 2001, p. 31). The fundamental argument of the book is that any reasonable mathematical analysis of these data shows how the litany of gloom and doom has influenced the minds of people who have used the statistical data in the environmental debate, and as a result they could not help but produce pessimistic foresights.

Lomborg's work is a meta-research study that examines the mathematical quality of the work of other researchers within the field of environmental science, and we should welcome it as such. This is one point to make about *The Skeptical Environmentalist.* Another point to make is, however, that at the same time, it commits the same mistake as the promoters of the litany of doom. Having read the book, one can only put it back on the shelf feeling that the statistical treatment of the data as presented by Lomborg is at the least to some extent convincing. This has been seriously questioned by the UVVU (Udvalget Vedrørende Videnskabelig Uredelighed [DCSD— the Danish Committees on Scientific Dishonesty]) but their critique seems deficient, at least to some degree. If Lomborg's work has a serious flaw scientifically speaking, it is similar to the one he opposes. Lomborg substitutes the litany with the celebration of future development through further technological research. However understandable this may be, if one considers the litany of doomsday scenarios to be deeply rooted and always taken for granted in the media and research on these topics, Lomborg's work is not mathematically reflective on its own enterprise. His book does not thoroughly take up the limitations of the formal approach that has been applied through his analysis. It may be that Lomborg is right about there being a decrease in the number of species that are extinguished, but the numbers hide that, for example, the Bengalian tiger is threatened. This may be an animal of special importance to the self-understanding of humans on this planet, not to mention the ecological systems of which it is a part. And what ethical value is attributed to the hunted animals? What cultural significance do these tigers represent? These are questions about ethics, values—anthropocentric or otherwise—knowledge from many overlapping sciences etc. and the mathematical analyses will always hide and often overlook its engagement with these concerns. Statistical analysis is created in a way that excludes the value of particular events, and can therefore always only be a partial story about the real state of the world. Lomborg knows he has made such preliminary ethical choices, but finds that his starting point is the only reasonable option, and he presents it on one page out of the 352 of the book. Lomborg positions himself in what he calls a human-centered view that, as he explains, focuses on the values attributed by humans to animals, plants, etc:

> This is naturally an approach that is basically selfish on the part of human beings. But in addition to being the most realistic description of the present form of decision-making it seems to me to be the only defensible one. Because what alternatives do we have? Should penguins have the right to vote? If not, who should be allowed to speak on their behalf? . . . It is also important to point out that this human-centered view does not automatically result in the neglect or elimination of many non-human life forms. Man is in so many and so obvious ways dependent on other life forms, and for this reason alone they will be preserved and their welfare appreciated. (Lomborg, 2001, p. 12)

Many would disagree with him on exactly these issues and there is no mentioning of the global political, cultural or economical conflicts and interests mentioned in this preliminary standpoint. Developing a well reflected and documented point of departure is where the real scientific debate should be focused and take place, and not exclusively—as has to a large extent been the case—with regard to mathematical technicalities.

What we more generally have in mind can be referred to in the subtitle of Lomborg's book: *Measuring the Real State of the World*. This formulation seems to presuppose that something could be called, not only the state of the world, but the "real state of the world," and that this state could be measured and objectively be decided upon once and for all. Here we find a similar assumption as the one expressed by Galileo, namely, that the essential aspects of the world can adequately be expressed in mathematical terms. In fact the idea is that mathematics is the unique descriptive tool, which captures the essential (physical) aspects of reality. In Galileo's terms what is essential are the primary sense experiences; these are the experiences that mathematics captures, thus leaving aside the secondary ones. Lomborg does not use this formulation, but in his analyses he (as well as those he criticises) concentrates on measurable aspects of the state of the world, and identifies these aspects as the real state of the world. In other words, when mathematics is brought into generate a description, the world is seen in a particular way, and when description becomes the basis for decision making, then mathematics is brought into action.

MATHEMATICAL TRANSFORMATIONS

What are we doing when we see the world through mathematics? And what kind of actions is connected to this way of seeing? By mathematics in action we refer to the actions that emerge as a result of taking a mathematical perspective on the world (or parts of it).[3] We can identify several aspects of seeing the world through the lens of mathematics, and how each of these aspects provides us with reasons for reflection.

Formalisation—Cutting off Parts of the Phenomenon

A troublesome aspect of a mathematical perspective of the world is presented by the Danish philosopher K. E. Løgstrup. In the third part, *Source and Surroundings— Reflections on History and Nature* of his four-volume work *Metaphysics* (first published between 1976 and 1983), Løgstrup suggests that we are always faced with the choice of interpreting the world as a causally governed system or as a phenomenological experience. One could understand this choice as a choice between attending to the primary qualities of objects in the world or the secondary qualities. It is Løgstrup's thesis that if we limit ourselves to studying causal relationships between primary qualities of objects we cut ourselves off from any human understanding of the universe as our source of existence. What is important to consider is to what extent the incompleteness of primary qualities, or the causally governed system as the basis for describing the world, poses a significant limitation to our understanding of the world. According to Galileo and to Modern Science there is no "limitation" connected to this incompleteness; the primary qualities can grasp what is essential to the world, scientifically speaking at least.

Løgstrup on the other hand finds that there are very serious limitations to what can be concluded about the world through studies of causal systems. When we use formal language as a means of describing a certain phenomenon in life, we cut off part of reality, that is, those aspects which cannot be captured within the conception of the primary qualities. We simplify matters within this field of vision *in order to* make causal judgements about it. In fact a formal description of the world will concentrate on those physical aspects of the world which make it appear like a causal system. This is not without reason. Science seeks to find explanation for what has happened or is happening, (that is finding the cause for the effect that is being observed) and use this to project or predict what is likely to happen if the cause remains. With this understanding, science can provide ways of thinking about what could be altered to the causal factors to diminish, increase or in some other way alter the effects in a desirable way. This is what the modern scientific enterprise is all about and it is one very important way for us to gain knowledge about our surroundings. Galileo emphasised this and in this way he opened the space for Modern Science. Alternatively one could try to retain the complex, the ambiguous and the paradoxical, as it is done in the extreme in the different forms of art and, to some degree, in the scholarship in the humanities. Although the traditions of Modern Science define science as the pursuit of knowledge through the use of formal language, we must always bear in mind that we have, in Løgstrup's phenomenological conception, cut ourselves off from seeing some of the attributes of a given phenomenon.

The exclusion of parts of reality through a description reflects the linguistic tool used for the description. In order to clarify this one can consider that natural language is the generative basis of formal language. Formal language is a derivative of natural language, which draws attention to certain aspects of natural language. We should not interpret formal language as the opposite of natural language as for instance Frege and Russell had done in their efforts to construct a scientific language separated and secluded from natural language. Instead we should consider formal language as a subset of natural language—natural language can in principle express what formal language can express, but in addition, it can do a whole lot more. Therefore we always have to be aware that formulating a description in a formal language means leaving out parts of the phenomena. A formal description is a highly selective description.

In this formulation, however, we might already have adhered to some assumptions of Modern Science which can be questioned. We talk about *reality*, and of language as *describing* this reality. This might well be a problematic formulation. The word description can be problematic, as this somehow assumes a form of "picture theory" of language, which views language as reproducing a mirror image of aspects of reality. And certainly the notion of reality has to be considered too: what can we assume when we make reference to "reality"? We assume a distinction but also a relationship between reality and its formulation through language. It makes more sense to talk about language and reality as two partly overlapping entities, which can be interacting in much more complex ways than indicated by a notion of description. Language can inscribe values into phenomena; it can form or categorise and restructure reality. And this also is true when the language is mathematics. We therefore need to reconsider the conception of a mathematical description of reality.

Systematisation and Inscription

Let us now assume that mathematics has been applied to describe some aspect of reality. The world is not taken in its phenomenological complexity, but is instead reduced to its primary qualities in the formal language of mathematics. We shall call this process in the sciences *formal reduction.*

What does this formal reduction mean for the way we *see* the world? In fact which world are we seeing? As a first step we can say that one is reducing a perceived phenomenon that involves both primary and secondary qualities into one that only reveals primary qualities.

But the reduction of a phenomenon does not stop here. After deciding upon a formal representation of the world's primary qualities there are still infinitely many possibilities for constructing a mathematical description of the phenomenon. The reduction of the world into measurable primary qualities has to be continued further. There are different mathematical lenses that can be put into use. One is about the size of things you are interested in describing. Are they nano-sized or are they cell-sized or perhaps planet-sized? On top of this come decisions about what causal elements needs to be added to the description. If we are dealing with the movement of a billiard ball we can reduce a description of the phenomenon by claiming that the surface is perfectly flat in some sense (which is never the case in reality), that the ball is perfectly round (which is also not the case), that the billiard table is a closed physical system (which is very far from being the case) etc. In conclusion we have to admit that from the phenomenon we have experienced—which can be far more complex than a billiard ball rolling across the table—we are after the formal reduction and the additional *system reduction* providing a description that has very little to do with the original phenomenon.

Hence, we have illustrated at least to stages of transformation in any mathematical description of a given phenomenon in the world. But yet another type of transformation takes place when we are constructing a mathematical description. It concerns the interests that are implicit in a given description of anything. These interests reflect the purposes that the creator of the description has in producing the description. Do they want to use a differential equation to produce a picture of an idealised physical phenomenon? Do they want to produce a table of numbers to show how prosperous a society is? Do they want to present a graph to highlight the inequities in the workload in an organisational unit? We shall call this third level of transformation an *inscription*, because it refers to the idea that certain decisions, values, intentions, interests, ideologies and priorities are built-in components of any mathematical description of the world (see Skovsmose, Yasukawa, & Christensen, forthcoming).

We have located three forms of transformations in relation to a mathematical description of a given phenomenon in the world; the formal reduction, the system reduction and the inscription. We may not even after these three forms of transformation have fully constructed a mathematical model as such. To obtain this we would still need to establish causal relations between elements, parameters, variables etc. in our mathematical description. In total we shall in what follows call the entire process of mathematisation the *mathematical transformation*. However, it seems quite clear that we have to be aware that what is normally called a mathematical description of a phenomenon cannot really be taken to be a predetermined map of the relations between things in the world. A mathematical description is something you subscribe to or do not subscribe to on the basis of careful examination of the values, limitation, priorities etc. of the transformations.

Mathematically Based Prescriptions

We now have an idea about what it could mean to make decisions based on mathematical descriptions, and that such descriptions include extensive transformations of phenomena. Once a mathematical transformation of a real phenomenon is made, then the transformed phenomena itself exists and takes on a life of its own. The formal reduction ensures that only particular aspects of the phenomenon become included in the descriptions and the systemic reduction ensure that a certain type of connections is established between the described entities. Elsewhere, we have introduced the notion of a mathematically scripted world (see Skovsmose et al., forthcoming). In this work we discuss how the mathematical script is used to prescribe certain actions, including decisions. People will then have a "choice" of subscribing to what has been prescribed as actions to take. In many cases, however, the subscription is so pervasive that the script has the appearance of conscripting certain actions. What is often overlooked is that like any script written by mathematics or natural language, mathematical scripts have inscribed in them certain ideologies and values. The level of subscription to the prescriptions may give the illusion that mathematical descriptions are value free, when in fact there is always inscribed into them particular values and interests. And very often the most fundamental value relating to our theme here is inscribed in the mathematical script as such— that the world is best understood through its primary mathematisable qualities. But this value is taken for granted and so not a subject of reflection when actions and decisions are prescribed on the basis of mathematical transformations. There may be arguments about the accuracy of the numbers, the number of variables that were used and in what way, but not "why use (only) numbers and quantifiable variables" in the first place.

Decisions turn into actions, and mathematics becomes part of reality. The actions taken as prescribed by a mathematical script are acted out in a complex reality, but they might only be justified within the world of the mathematical transformation of the phenomenon. Through a mathematical script one can for example formulate certain standards, for instance, concerning the "acceptable" degree of pollution of drinking water. Such standards are established through a mathematical modelling process. However, when first established such standards are not only part of a model, and represent certain prescriptions; they in fact create a new reality. The standards make part of the risk structures, which constitute our life conditions. Our health could be protected behind such standards.

CONCLUSIONS

Mathematics brought into action is a powerful resource for confining the breadth and possibilities of criticisms of decisions. In order to challenge mathematically formulated actions, one is possibly expected to challenge it within the internal world created by mathematics. This is not always possible, particularly when the criticism is about how the mathematically formulated world is interacting with parts of the world that the mathematical transformation left out. But leaving things out, allows for the mathematisation in the first place.

Returning to Lomborg's environmental book, an interesting feature was the measure of attention it received in the public debate. Lomborg is backed by a considerable public consencience, because he is not afraid to talk about what is right and what is wrong in the environmental debate. This attracts the media and influences the public opinion. It displays a feeling of lack in the public debate of science making a clear cut comment on what is right and what is wrong in for example the environmental debate. Science should present a given case to the public as complex, undecided, based on limited knowledge and so on, if this is actually the state of our knowledge in that particular field of investigation. We should be thankful to scientists when this is how they reply to our questions. But Lomborg's crusade against the litany of doomsday very convincingly showed us how mathematical transformations are also embedded in power struggles—in this case about what path to proceed along in environmental issues. The debate was an important illustration of the need for people involved in mathematical modelling to be reflective of the diversity of approaches that can be pursued in the study of a given phenomenon—in this case the state of the world!

Our intention here has not been to criticize the use of formalisation in science—we cannot do without formalisation and especially not in science as we know it today. Our concern is in the blurring of the (mathematical) model with reality itself. This blurring has a long history, starting at least from the conception of science in modernity that the world can be spanned by our formalism and that this world is the unedited, uncut and entire world. We could go on and talk about the blind spots of mathematical transformations. These blind spots represent what is left out in order to perform a scientific formal representation of a phenomenon, and it is not visible from this formal framework itself. In other words, we shape the world to our mathematical approach in order to talk scientifically about it.

In conclusion it seems important to consider the impact of our life world becoming shaped by formal mathematical approaches. We are in the process of formalising our cultural environment—the world as we experience it—so that we increasingly experience our life world as formalised. We are not merely describing the world through mathematics but rather transforming it into categories accessible through and computable in mathematics. Only when we become aware of this transformation produced by a mathematically scripted world, can we retain the possibility of a radical critique. As long as mathematics is churning out consistent answers, there is no easily accessible space for reasonable critique and formalisation measures will continue to dominate the construction of our life world.

ACKNOWLEDGEMENT

This article has been previously published in the *Alexandria: Journal of Science and Technology Education 1*(1), 77–90. It has been reprinted with permission from the journal and the authors.

NOTES

1. See, for instance, Delanty, G. and P. Strydon (eds.) (2003).
2. The Worldwatch Institute has been publishing *State of the World* annually since 1984.
3. For a discussion of mathematics in action or the formatting power of mathematics see also Skovsmose (2005); Skovsmose and Yasukawa (2009); and Skovsmose et al. (forthcoming).

REFERENCES

Crosby, A.W. (1997). *The measure of reality: Quantification and Western society 1200–1600.* Cambridge: Cambridge University Press.

Delanty, G., & Strydon, P. (Eds.) (2003). *Philosophies of social science: The classic and contemporary readings.* Maidenhead: Open University Press.

Løgstrup, K. E. (1995). *Ophav og omgivelse—Betragtninger over historie og natur, Metafysik III.* Copenhagen: Nordisk Forlag A.S.

Lomborg, B. (2001). *The skeptical environmentalis—Measuring the real state of the world.* Cambridge: Cambridge University Press.

Skovsmose, O. (2005). *Travelling Through Education: Uncertainty, Mathematics, Responsibility.* Rotterdam: Sense.

Skovsmose, O and Yasukawa, K. (2009). Formatting power of 'Mathematics in a Package': A challenge for social theorising? In P. Ernest, B. Greer and B. Sriraman (Eds.), *Critical Issues in Mathematics Education* (255–281). Charlotte, NC: Information Age.

Skovskmosem O., Yasukawa, K. & Chrstensen, O. (in preparation). *Scripting the World in Mathematics,* forthcoming.

CHAPTER 8

HUMOR IN E(A)RNEST

Stephen I. Brown
USA

INTRODUCTION

Paul Ernest has laid out an ambitious agenda in seeking and applying the socially constructed underpinnings of a field that has been perceived by many as strictly axiomatic, logical, and rule bound. In addition to his other writings, many new questions, debates, and analyses about the nature of teaching, learning and doing mathematics—and about the nature of mathematics itself—emerged as a result of his editing POME, the *Journal of Philosophy of Mathematics Education.*

This article picks up on and expands upon the concluding theme in a recent POME essay of mine—that of humor. In that essay I stood on its head what was perceived by western educators to be shortcomings in the mathematical thinking of members of the Kpelle tribe (Brown, 2007). There I sought to understand the strengths of grasping the world through their eyes—story-telling (and especially gaining kudos by exaggerating stories).[1]

I have striven to make this Festschrift essay accessible to TCMITS (*The Celebrated Man in the Street;* derived from the title of a book by Lillian Lieber, 1942) as well as to TSPITS [*The Scholar Person in The Stratosphere,* my modern day extension of Lieber's clever acronym]. While making use of philosophical concepts, I have minimized the use of technical jargon. Though

---

*Relatively and Philosophically E*ª*rnest,* pages 95–126
Copyright © 2009 by Information Age Publishing

for TCMITS, I have selected mathematical examples that require only a modest amount of prior knowledge, they most likely will be among the arsenal (no more militaristic metaphors) if not part of the treasure chests of TSPITS.

"BLACK" HUMOR: AN ENTICEMENT

How might mathematical thinking be related to humor? This question itself might seem a tad funny (peculiar?). For most people, the association of mathematics with humor is absurd. To seek connections, it will be helpful to explore what it is that theorists of humor have to say about the nature of humor.

As an entrée, I recount the beginning of one of the cleverest lectures I ever heard. It was given by Max Black (at the time both a Cornell philosopher and alive), whose philosophical writings spanned topics such as language, art, foundations of mathematics and science. The lecture began with a slight variation of the following two jokes:[2]

1. Mr. Smith, an American in Paris who knew very little French, was eating in an exclusive French restaurant. He noticed that there was a fly in his soup. He called the waiter over, took out his French–English dictionary and pointed at the fly in the soup:

   Mr. Smith: Garçon, *un mouche,* un *mouche!*
   Garçon: Non monsieur, *une* mouche, *une* mouche.
   Mr. Smith: What fantastic eyesight!

2. Mr. Gould meets an old friend, Mr. Brown, in the street. They had not seen each other for many years. The following conversation ensues:

   Mr. Gould: So, how are you?
   Mr. Brown: Awful.
   Mr. Gould: Why?
   Mr. Brown: I just came back from the doctor with my wife and found out that she has an incurable disease and has only six months to live.
   Mr. Gould: It could be worse.
   Mr. Brown: And also I was just in an automobile accident and my brand new car was demolished and I had no collision insurance.
   Mr. Gould: It could be worse.

Mr. Brown: I got a phone call last week from my son, who has been married for twenty five years and he tells me that his wife, whom I adore, and who has been wonderful to me and my wife, is planning to divorce him.

Mr. Gould: It could be worse.

Mr. Brown: My partner came to my house yesterday and told me that he has been cheating our customers for years. Someone has informed the government and we are being sued.

Mr. Gould: It could be worse.

Mr. Brown: What do you mean by telling me "It could be worse?" I tell you my wife is dying; my son is about to get a divorce; my car is destroyed, and I am about to become bankrupt. What could be worse?

Mr. Gould: It could be me!

It is not so much that the jokes were hilarious (though they were clever). Rather it was the intent of the jokes that made the lecture so clever. Max Black continued with about a dozen additional jokes. He then challenged the audience to come up with a single theory of humor that covered all of them. Before reading further, readers may wish to see what theories might explain the humor of these two jokes.

SOME THEORIES OF HUMOR

Mindful that joke telling and humor are not synonymous, Black was using this format to explore the question of whether or not humor can be reduced to one theory. His point was that it was not possible to do so. There are, in fact, a number of different theories of humor, and it remains an interesting question whether or not they can be consolidated into one.

Rather than resolving that issue, we shall review briefly a few of the well known theories of humor and will then explore the ways in which mathematical thinking shares significant attributes with one of them.

One of the earliest theories of humor is due to Aristotle, who wrote a lot more about tragedy than humor. The Greek conception of humor associates it with base instincts. It has to do with what is ugly but not disastrous. Derived in part from the above is a theory of humor that was popularized by Thomas Hobbes (1914/1651) in the *Leviathan*. His theory, known as "sudden glory" associates humor with a sense of superiority, a realization that one's station is superior to that of the person(s) who are seen as humorous.

Freud (1960), not surprisingly, had a theory of humor that spoke more about its function than its qualities. In his *Jokes and Their Relation to the Unconscious*, he claims that humor enables people to vent their aggressive and repressed sexual feelings.

Among the best known theories of humor were those espoused by Henri Bergson, Arthur Koestler and Molière. There are many variations of this perspective, but the connecting feature is that humor is associated with incongruity. Paulos (1980) in *Mathematics and Humor* adopts a variation of this theory in seeking connections with mathematics. He points out correctly that if incongruity is a necessary condition for humor, it surely is not sufficient, for there are many incongruous events, ideas, circumstances that are not at all funny. Paulos claims that in addition to *having* incongruity, humor requires (a) that the incongruity be noticed, (b) that it have a point, (c) that the emotional climate must be right.

Incongruity is a helpful umbrella within which to explore mathematical aspects of humor, but it is a large and somewhat amorphous one. In the following section, I will seek to explore the scope of that concept through a number of examples. I am interested in feedback from readers to see how much I have expanded the umbrella into a Procrustean bed. I should warn the reader, however, that I have knowingly but reluctantly included the sub-category of *self-referentiality* in the following section. I plead guilty to having set that category in such a bed, for though it is possible to have it slumber there on the grounds that it wears a bonnet of incongruity; it smothers all the other categories. Since it does have a life of its own. I have compensated for my misdeed by giving it a very healthy berth. If readers find its location troublesome, I will not be offended if they choose to cut and paste to bequeath it a honored independent section of its own.

KINDS OF INCONGRUITY

As we explore several different sorts of incongruity in mathematical thinking, it will become clear that they are intertwined. Such is the result of allowing specific examples to take on a wife (sic) of their own. In selecting these categories of incongruity, my intent is to be illustrative rather than exhaustive and to use them to seek out important aspects of the mathematical experience. Surely incongruity can be found in other categories that relate humor to mathematics. Though we have mentioned "figure/ground" switch and "proof by contradiction," for example, we have not highlighted them as sub-categories of their own. Others are described in Brown (2001) and Brown & Walter (1983, 1986, 1990, 2005).

Breaking Expectations: Changing Anticipated Set

One sort of incongruity is that which minimizes laborious calculation in favor of insightful inversion. Though the story of young Gauss (1777–1855), the "Prince of mathematicians," is perhaps apocryphal, it illustrates an attitude of honoring genius early in one's career.

His elementary school teacher, wanting relief from working with her pupils in order to perform some necessary task (like collecting milk money) supposedly asked students to find the sum of the numbers from one to one hundred. Gauss is alleged to have gotten the answer in a very short time, not by laboriously adding the numbers, as his teacher had anticipated, but rather by engaging in an act of clever problem solving—something akin to a gestalt-like shift in perspective. While other students saw the fact that each addend was one more than its predecessor, he saw a connection that made the problem curl up in embarrassment, as is illustrated in Figure 8.1.

Seeing that what was taken away from numbers at the end of the sequences was compensated for by a comparable increase of numbers at the beginning, he paired elements at both ends so that there was a constant sum of 101 repeated over and over again. Thus he perceived an important constant among the significant variety of components of the problem—a constant that enabled him to figure out the answer (fifty pairs of 101) in his head.

It was not only that Gauss solved a problem so the story goes, that was perceived by his classmates to be a virtual nightmare. More importantly he created a scheme, which in some sense got to the (or perhaps "an") essence of the problem.[3] He reduced something unmanageable to something manageable. He observed similarities of structure among a morass of disparate "facts." He was not distracted by intrusions that may have been true but that did not bear on the essential features of the problem.

So, at bottom, what Gauss did was to see a problem that had implicitly been viewed through one set of lenses (involving laborious calculations) and created an alternative set of expectations (involving an underlying structure that had not been taught to him).

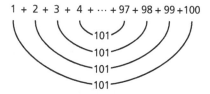

$$1 + 2 + 3 + 4 + \cdots + 97 + 98 + 99 + 100$$

Figure 8.1 Depicting Gauss's alleged insight for adding numbers from 1 to 100.

For an example that has the potential to by-pass an even greater amount of manipulation, Paulus (1980) offers the following:

> A checkerboard has sixty-four squares. You can imagine covering those sixty-four squares with thirty-two dominoes. Now remove two diagonally opposite corners of the board. It would seem that thirty-one squares could cover the new board. Will they? (p. 13, paraphrased)

In the sections entitled, "Reflections on Incongruity" and "Punch Lines," we will sharpen what is involved in the kind of thinking described in this category, and will suggest broader ways of conceiving of "talent" and "genius."

Process and Product

One of the most elegant proofs in mathematics, and one that has more to appreciate each time I rethink it, is Euclid's proof of the infinity of primes. The proof must have been inspired by an enormous sense of doubt and wonder. Once we define a prime number (in the set of natural numbers) as a number with exactly two different factors, it is easy to enumerate some of them. So, the first few are: 2, 3, 5, 7, 11, 13.... Having listed them, it becomes interesting to wonder how many there are. How far down the line might we have to explore them before we reach the end?

The answer was forthcoming over two thousand years ago (around 300 BC) by Euclid. His work—an early attempt to organize mathematics as a collection of formal proofs—consists of 465 propositions that are based upon only five axioms (like "Things that are equal to the same thing are equal to each other") and five postulates (like "It is possible to draw a straight line from any point to any other point"). In Euclid's *Elements*, we find thirteen "books," each one dealing with a different theme. (See Heath, 1956) Most of high school geometry is derived from the propositions in this collection.

The gist of Euclid's proof that there are in fact an infinite number of primes, consists of the following proof by contradiction (from Book IX, proposition 20):

He created a new number out of any supposedly finite list.

So consider the list: 2, 3, 5, 7, 11, 13, ..., p, where p is the last prime. Euclid creates

$$N = (2 \cdot 3 \cdot 5 \cdot 7 \cdot 11 \cdot 13 \ldots \cdot p) + 1.$$

The production of such an N *guarantees* (through a brief argument we omit here) that there must be some new prime other than those listed.

We could just follow his proof and not be stunned by it. What is stunning is not only that it is brief and that it appears to come from out of left field (and not something that would come to mind in a short time by most of us), but that it has an element of disguised incongruity built in that broaches on the ironic.

The irony is in the creation of N itself, for if we think of N not as a result but as a process (over time perhaps), and *retreat* one step in its construction (deleting the " + 1" at the end), we bump up against one of the most non-prime numbers we can meet. That is with the deletion of " + 1" the number is divisible by everything in sight. Just adding 1—the smallest available counting number possible—to that mega non-prime number totally reverses that condition!

In order to appreciate the aesthetics of the incongruity, we need to pay attention to a different sort of emphasis than we customarily focus upon if we are primarily concerned with the logic of proof. What is so astounding is not only that Euclid was able to come with such a brief and elegant proof, but that he did so by entertaining even for a millisecond (and not discarding his find) a humongous number of composites along the way.

Instead of anticipating what follows and waiting for a proof to continue, we might profit from holding onto that observation as an aesthetically pleasing and unanticipated act of construction—and one that stops us dead in our tracks.

Inversion

When we are examining a statement that is so obvious that there is little to do to verify it, what are we inclined to do? There are all sorts of things we might do. Let us look at one and then think about how we might proceed.

Standing an idea on its head is one of the most powerful generators of humor. What is so funny about Max Black's second joke. We have been focusing in this joke on the sorry state of Mr. Brown's health, and we expect him to continue to tell us more about his troubles.

Instead what happens? Mr. Gould informs Mr. Brown that the object could be him rather than Mr. Brown.

Making use of all forms of inversion in mathematics has the power not only to seek proofs and to solve problems, but to generate questions to ask in the first place. It sensitizes us not only to pay closer attention to what may appear to be trivial but to marvel at what is just beneath the surface.

One powerful example of the inversion of a trivial observation may have accounted for a brilliant realization in the mid eighteenth century by Christian Goldbach.

**TABLE 8.1   Expressing Even Numbers
in N as the Sum of Two Odd Numbers**

$$4 = 2 + 2$$
$$6 = 3 + 3$$
$$8 = 5 + 3$$
$$10 = 5 + 5$$
$$12 = 5 + 7$$
$$14 = 7 + 7$$

Since all primes greater than 2 must be odd, it is easy to see that if you add two primes (excluding 2), you will always arrive at an even number. That is too trivial an observation to get excited about. An interesting follow up on the nature of the statement however leads to something that has puzzled mathematicians for a very long time. That is, suppose that instead of examining the obvious statement, we look at its converse. The making of such an innocent appearing "tweak" is one way of hypothesizing how Goldbach came up with the following conjecture, "Every even number greater than 2 can be expressed as the sum of two primes."

There are some simple instances that seem to support the conjecture, as indicated in Table 8.1.

In some cases, there are two different ways of finding sums of pairs of primes for a given even number. In the above case, for example, we find that in addition to what appears in Table 8.1, we have:

$$10 = 5 + 3 = 3 + 7$$

$$14 = 7 + 7 = 3 + 11$$

Goldbach's concern though was not with *uniqueness* of pairs, but rather with the *existence* of a pair for every odd number.

It turns out that very little headway had been made with this conjecture for over two centuries. Then in 1931, another Russian mathematician, L. G. Schnirelman created the first crack in an effort to establish that it might be true. To appreciate what he proved, recall that Goldbach wanted to establish that at least one *pair* of primes could be produced for any given even number.

What did Schnirelman do? Given that Goldbach was hoping to show that at least one *pair* could always be found, what would be a really funny answer to how many primes you might have to add *before* achieving the desired result: a specified even number? Perhaps ten would be funny. Maybe 100, or perhaps 1000 would create a chuckle. Would the need to add 300,000 natural numbers instead of a pair create a belly laugh? That was Schnirelman's contribution to the problem.[4] He showed that given any even number, we

Figure 8.2 Visual reversal of figure and ground.

can find at most 300,000 primes that must be added in order to achieve it! (See Dunham, W. [1990].)

While the Goldbach example focuses on inversion from the point of view of converses, there are many different forms of inversion. One that has particular visual appeal is that of reversing figure and ground. In Figure 8.2, what do you see? Depending upon what you select as figure and what as ground, you will see either a vase or two people facing each other.

Another form of reversal that has powerful generative potential involves an inversion of axiomatic properties.

There are several fundamental properties of arithmetic from which others can be derived.

Looking, for example, at the natural numbers, we can observe that among the critical properties for addition and multiplication are the commutative and the associative properties. Thus:

$$a + b = b + a; \ a \cdot b = b \cdot a \qquad \text{(commutative properties)}$$
$$a + (b + c) = (a + b) + c; \ a \cdot (b \cdot c) = (a \cdot b) \cdot c \qquad \text{(associative properties)}$$

In each of the equalities above, one operation is involved. If we switch addition for multiplication (and vice versa) in any one of the equations, we end up with an analogous property that is also holds for all numbers. The commutative and associative properties are duals of each other.

There is one property, however, that mixes both addition and multiplication in an obvious way: the distributive property:

$$a \cdot (b + c) = (a \cdot b) + (a \cdot c).$$

What happens if in this property, we switch addition and multiplication?

We end up with:

$$a + (b \cdot c) = (a + b) \cdot (a + c)$$

Surely this new property does *not* hold for all natural numbers.

As a start we might wonder when the modified distributive property—the "dual" of the original one—holds. Also, are there systems (other than the natural numbers) for which it might hold with the same sense of abandon as the other properties? Actually this act of inversion inspired me to ask an intriguing meta-mathematical question: one about the system itself. That is, I wondered—if there was a way of knowing before any actual attempts to do so—whether theorems in the system of natural numbers *could* be proven without using the distributive property (see Brown [1969]; Brown & Walter, [2005, pp. 102–104]).

"Existential" Thinking

While the work of Goldbach and Schnirelman is interesting from the point of view of creating an unexpected area to explore, it would most likely *not* have been among the first problems people would have investigated once they discovered that there are an infinite number of primes. They most likely would be more interested in finding out such matters as how primes cluster or producing a formula that would generate them.

Well known mathematicians, including Fermat and Mersenne in the seventeenth century, and Euler in the eighteenth century, devoted years of their lives attempting to make headway on the problem. (For an elaboration of the number theoretical contributions of Fermat, Mersenne and Euler, see Brown [1978, 1991]).

Perhaps the most interesting way to understand the depth of the problem, and ultimately the humor involved, is for us to whittle away at the question being asked. If we know that the number of primes is infinite, then the simple formula "n" for all natural number n would do the trick. Well . . . yes, in a sense. The formula would generate all the primes and even do so in sequence, but at a price, i.e., it would generate all the non-primes along the way as well.

The most ambitious request would be for a formula that generates all the primes in sequence and none of the composites. Short of that, we might hope for a formula that would generate all the primes and no composites even if the sequence is destroyed. Nibbling away even further at expectations, we might hope for a formula that would generate only primes, even if many were left out of the sequence. Reducing our expectations even further, we might have to settle for a formula that would generate an infinite number of primes even if an occasional (perhaps an infinite number) composite were viewed as a necessary evil.

In 1947, the problem was solved at a middling level of desire. More precisely, M. H. Mills came up with a formula that would generate primes and only primes (forever) for every substitution of n in a simple formula, though some of the primes might be left out. What was the formula?

The good news is found in the apparent simplicity of the formula. The bad news, however, is wrapped up with its concomitant humor. The formula is:

$$[A\,3^n]$$

That is, there some fixed number A such that each time n is plugged into the above formula, we will be handed a prime number on a silver platter.

The meaning of the "[ ]" is simple. [2.57] = 2; [4.2] = 4; [1.9999] = 1. [x] stands for the greatest integer less than or equal to x.

What a find! Yes, but of course in order to make proper use of the formula, we have to know the value of A. What is it? Answer: Though much effort went into the production of the above formula, nothing is revealed about the nature of A. A might be some real number (e.g., the square root of 2) between 1 and 5; or it could be a number that is larger than the number of stars in the sky. (For further elaboration, see Brown, 1978, 1991, pp. 17–20.)

Why is that funny? The incongruity is between what we might expect in the way of an answer to some numerical question in our daily lives verses what we seem to be satisfied with here. If you wanted to know when expected guests would be arriving for dinner, and someone told you that there was a definite time they would arrive, but they could not tell you if it might be within the hour or within the decade, it would of course seem ludicrous.

Well, perhaps we were handed the prime number not on a silver platter, but on a stainless steel ("steal," more accurately) one.

Assertions that *there exists* something (like some number A having a desired property), even though we may have no idea of its whereabouts, is a kind of statement that is rampant in mathematics, but it is used quite sparingly in most other contexts. We do know of such real-world "existential" statements however. For example, we know (as an empirical inductively arrived at fact) that there is a precise moment at which we will die. That realization may in fact direct the way we live our lives, though it is something we tend for the most part to repress. Without greater specificity, however, it is not the kind of statement that will have much of an impact on how we live in any fine-grained way.

Paradox

The following is a variation of a problem (designed especially for readers as they read this essay) that allegedly caught Einstein's fancy. Here is a version of it:

> Three people reading this essay become tired of it all and decide to go to a nearby carnival for recreation. They take a ski lift together and they notice that it moves at a steady pace and that there are markers along the way. One person keeps track of the markers and another keeps track of the time. They notice that the lift travels one mile every four minutes.

> They reach the top, and looking for "adventure of a new and different kind," they decide to ask the person who controls the ride to send them back down again at a constant speed so that the average speed of the entire trip is twice that of the speed on the way up. What rate should the downward journey be set at, so that the group will be pleased that they will not feel the urge to complete this essay when they return?[5]

Before doing any calculation, what would you guess would be a reasonable answer?

Rather than offering a solution, we suggest, if you have not thought about this before, that you get together with two of your potential ski lift friends and come up with your solutions independently. Compare them. If you have different answers, see if you can figure out how any of them that are different from yours may have been calculated? Rather than my spoiling it for the reader, it should suffice to suggest that most intuitive answers are incompatible with various forms of calculation. Until one changes perspective on this problem, it seems hard to adopt a strategy that would invite a non-paradoxical solution.

In some paradoxes, the incongruity or logical contradiction is apparent immediately. We know that something is wrong before examining alternative approaches.

An example of this is the famous Zeno paradox:

> One of the most famous of Zeno's paradoxes involves Achilles and the tortoise, who are going to run a race. Achilles, being confident of victory, gives the tortoise a head start. Zeno supposedly proves that Achilles can never overtake the tortoise. Why?

> Before Achilles can overtake the tortoise, he must first run to point X, where the tortoise started. But then the tortoise has crawled to point Y. Now Achilles must run to point Y. But the tortoise has gone to point Z, and so forth. Achilles is stuck in a situation in which he gets closer and closer to the tortoise, but never catches him.

In other cases, it is necessary to explore competing explanations before the paradox becomes apparent. That is the case in the ski jump paradox.

Self-Referentiality

Yet another perspective on the relation of humor to mathematics has to do with the theme of self-referentiality. It is evoked, loosely speaking, whenever an object refers to or calls upon itself. I experienced the concept most dramatically many years ago when my daughter Sharon, who was four years old, asked me, "If God created everything, then who created God? You tell me that!" Though some might dismiss rather than answer the question, others might assume that the question is both meaningful and difficult even if impossible to answer.[6] (Though it would be getting ahead of our story, if Sharon were a few years older, I can think of a mathematical analogy that would point to a surprising twist in response to her question. It would shed light on the nature of questions themselves[7]).

In his recent book, *I Am a Strange Loop* Douglas Hofstadter (2007) follows up on a theme that was central in his earlier prize winning *Gödel, Escher, Bach: An Eternal Golden Braid*, (Hofstadter, 1979). Once again he envelops himself with the concept of self-referentiality both as a theme to propel his story and as a concept to analyze.[8] Combining personal reflection and philosophical analysis throughout, he points out a pervasive insight. Before getting to the zinger, we begin with his story:

> When I was twelve, a deep shadow fell over our family. My parents, as well as my seven-year-old sister Laura and 1, faced the harsh reality that the youngest child in our family, Molly, then only three years old, had something terribly wrong with her. No one knew what it was, but Molly wasn't able to understand language or to speak (nor is she to this day, and we never did find out why). She moved through the world with ease, even with charm and grace, but she used no words at all. It was so sad.
>
> For years, our parents explored every avenue imaginable...and their quest for a cure or at least some kind of explanation grew ever more desperate. My own anguished thinking about Molly's plight...gave me the impetus to read a couple of lay-level books about the human brain. Doing so had a huge impact on my life, since it forced me to consider, for the first time, the physical basis of consciousness and of being—or of having an "I" which I found disorienting, dizzying, and profoundly eerie. (p. xi)

Motivated by this gripping event in his life, Hofstadter discusses how he was taken over by an analysis of the meaning of "I." He explores what makes a person a person, and what it means to have a mind, to have consciousness, and to explore whether or not other organisms might have it—and more generally what it means to be human. Appreciating the force of self-referential thinking, he delights in the fact that he (an "I") is in search of "I."

So far, so serious. How is humor expressed within this context? One of the delightful qualities of the concept of self-referentiality is that it frequently is applied to events that generate comments which are on the cusp of sarcasm meaninglessness and incongruity. Groucho Marx, who is reputed to have exhibited one of the most famous self-referential comments when he made the whimsical remark, "I would never join a club that would accept me as a member."

Not only verbal jokes, but the vase/faces of Figure 8.2, and the famous drawing of hands in Escher's work (Figure 8.3), indicate some well known visual images that evoke self-referentiality.

Paulos (1980) has a cartoon of people sitting in a train beneath an advertisement which reads, "Want to Learn to Read?" It then gives a telephone number for further information. The logic of the cartoon is both apparent and "grabby."

How does mathematics partake in this adventure? It turns out that the concept of self-referentiality or its near relatives flourish in quite diverse fields, from mathematical logic to number theory, from chaos theory to combinatorics, from geometry to computer science, from literature to films as well.

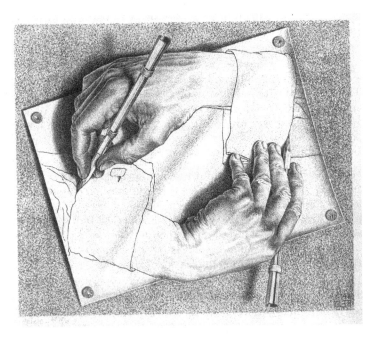

Figure 8.3 Hand drawing the hand.

A simple example from logic itself would be:

1. "This sentence is between five and ten words long."
2. "This sentince has three erors."
3. "This sentence is false"

Though it refers to itself, sentence (1) above is both straightforward and true. Sentence (2) however, is delightful and not only exemplifies self-referentiality, but itself verges on the humorous. It looks as if there are only two errors in the sentence, both of them spelling errors. So what is the third error? Look again at the sentence. The third error talks not about the individual constituents of the sentence, but of the sentence as a whole. That is, the claim is made that the sentence has three errors, but that claim about the sentence itself is wrong, for it has only two errors. Therefore, the third error in (2) above is the claim that the sentence has three errors!

Sentence (3), however, is more interesting. Its humor resides in its playfulness: a kind of playfulness that is depicted in the cartoon of Paulos that we described above. Actually this kind of analysis had a particularly significant impact on the foundations of mathematics. In mathematics, sets were taken to be powerful, fundamental, and unproblematic building blocks of mathematics from the nineteenth through the early part of the twentieth century. In 1908, Bertrand Russell brought the theory of sets, to a screeching halt by showing that it was not so easy to speak of sets as a clearly defined concept. He essentially created an innocently appearing new set out of two old ones, which led to some quite problematic consequences.[9]

The use of this concept enabled Kurt Gödel in the early 1930s to engage in something every bit as devastating as Russell's finding. The analysis crushed the mathematician's optimism that the field could be viewed as the last bastion to defend the "Work Ethic." Through the use of self-referentiality on a scheme involving the natural numbers, he showed that intelligence and hard work cannot confront inherent limitations of mathematical provability.

He set an agenda to demonstrate that mathematics as a field could not lead to inconsistencies. He also hoped to establish that we could create descriptions of mathematical systems that were complete in the sense that one would eventually be able to prove or disprove any well formed statement in mathematical systems that were at least of the order of interest and complexity of arithmetic. He hoped thus to show that if any statement could not be proven to be true or false at any particular moment in time by use of the well established axioms, then this was merely an indication of human frailty. Proving or disproving conjectures would then be a matter of commitment of time and an investment of cleverness of successive generations.

Not only was the longing for a demonstration of absolute consistency found to be a pipe dream, but he also showed that in a system as ostensibly mild as the set of natural numbers, it is impossible to create a formal structure that would enable one to prove or disprove every statement belonging to it. That is, there must exist undecidable statements in any such system—statements that are true but unprovable as such.

How to interpret his findings and exactly what they might mean for a formal view of mathematics is a problem that philosophers of mathematics have grappled with ever since. It is clear however, that Gödel has in some sense hoisted rigor and hung it on its own petard.[10]

One of the most recent curriculum concepts associated with self-referentiality is due to Benoît Mandelbrot. Born in Warsaw in 1924, he was a refugee from Nazi Germany, a dabbler in many applied fields and a self-made mathematician. Coining the word "fractal" in 1975 for unusual shapes he discovered in nature, he contributed significantly to the newly emerging field of chaos theory—seeking regularity in irregularity. Mandelbrot's (1999) theory applies to fields as diverse as geometry, coastlines, and the stock market.

He confronted our common sense belief that as we take supposedly more detailed measurements, we will reach some upper limit for the "true length" of the coast line of England, for example (if not the perimeter of a piece of *Almond Bread*: sic).

In fact this intuition is incorrect, though it would be true if the coastline were some Euclidean shape (an arc of a circle, for example). Then his method of adding smaller and smaller lengths of segments of straight lines would indeed converge to some limit. But Mandelbrot found that as the scale of measurement becomes smaller, the measured length of a coastline rises without limit. See Gleick (1987) for further discussion of this as an example of chaos theory.

An additional area of application of self-referentiality is in film making and literature. In Woody Allen's *Purple Rose of Cairo* (See Greenhut, 1985), one of the characters portrayed in the audience at the beginning of the movie not only loses herself in the movie, but she becomes a character in its ongoing plot. Even modern novels operate in such self-referential ways. In Raymond Federman's (1995) film *Smiles on Washington Square*, for example, the characters advance the plot by inventing each other. In other novels, characters not only decide to move the plot in a new direction, but they do so in conversation with the novelist.

The theme is a deep one and, like Hofstadter, the more I think about personal matters, the more I find myself thinking in self-referential ways (as many parenthetical remarks—including this one—will indicate). That is, as we indicated earlier with Hofstadter's quote, mind has the quality of asking itself "What is a mind?" Attempts at answering this question reveal unusual insights. As mentioned above, one of the most interesting aspects of the concept is that it appears to be on the cusp of rationality.

Sometimes it makes a lot of sense; sometimes it is confusing; sometimes it leads to surprising contradictions.

As a reminder that at least one popular variation of the concept is both central and problematic in mathematical thinking, consider mathematical induction as a form of proof.

One version of its abstract formulation is as follows:

Let $P_1, P_2, P_3, \ldots$ be a sequence of propositions. Then if:

If P_1 is true;
If P_n implies P_{n+1} for all n, then
all the propositions $P_1, P_2, P_3, \ldots$ are true.

A typical application of this form of proof might be to establish the validity of Gauss's famous discovery to which we alluded earlier. That is, suppose we want to show that:

$$1 + 2 + 3 + 4 + \ldots + n = [n \cdot (n + 1)] / 2$$

Without making use of Gauss's brilliant insight, we could do something a bit more "plodding" and make use of mathematical induction to establish the equality of the two sides of the above equation.[11]

Many people who have taught this form of proof have met interesting resistance by their students at early stages. They might actually do well to encourage resistance, especially if it appears that a form of brain-washing has set in. Even if students have no difficulty with the *technique* of establishing that P_n implies P_{n+1}, feeling that the reasoning is basically circular may linger.

Clever use of metaphors (such as thinking about mathematical induction as being like a collection of standing dominoes with the understanding that knocking any one of them down has the effect of knocking the one next to it, and so forth), may change students' reluctance to engage in mathematical induction and may even be persuasive. A curriculum which takes a look at the panoply of ideas that employs self-referential thinking, however may not only persuade them of the justification of the enterprise of mathematical induction, but may also enable them to see that it is reasonable to have doubts.[12]

Mathematics and humor are linked in their need for and use of a variety of self-referential schemes. What is particularly intriguing is that no matter how rigorously the self-referential schemes are developed, we frequently find that they wrap the object of their affection in a cloth that is simultaneously deep and fanciful. A number of the mathematical examples we have reviewed in this section entice and beckon us because they appear to come with a tag that says, "Should I take this seriously or not?" "Is someone just pulling my leg, or am I confronting a fundamental truth of the universe?"

REFLECTIONS ON INCONGRUITY

In an attempt to expose elements of mathematical thinking that are shared with humor, we have identified incongruity as an important domain to explore. Though incongruity is an important part of the mathematical scene from many points of view (especially that of generating new ideas), the use of and appreciation of incongruity does not *necessarily* exhibit humor in mathematics.

It sometimes does, however, and once we have begun to seek humor in mathematics, we will be able to identify many more instances than we had imagined.

There were several examples in the section above that did in fact indicate high humor in mathematical explorations. We observed it in the use of Mills' *existential* statement in coming up with a formula to generate primes, and also in Schnirelman's proof—in response to Goldbach's conjecture that every even number (greater than 2) can be expressed as the sum of two primes.

Schnirelman's creation of a proof that indicated that we would not need to add more than 300,000 primes when Goldbach was hoping to find a proof that two would do is hilarious. It is not that Schnirelman came up with a proof that 300,000 would be an upper limit, but rather that his demonstration was viewed as a form of progress in the first place. This is so despite the fact that successive approximations to the truth abound in all fields. Perhaps there's something about schooling that tends not to honor that realization in terms of student performance.

The incongruity between what is usually called for in "practical" situations and what is deemed to be appealing (and sometimes just a first approximation in refining a problem) in mathematics is frequently a source of high comedy and, such comedy is often rooted in "existential" mathematical proofs—especially when existence is left hanging off a cliff in the sense that we have no idea of the color of the cliff, its height, its texture and its surroundings . . . only that it is there and that it is powerful.

Even when humor is not involved directly in incongruity, it frequently does serve a function that is not unlike a role that humor plays. It often provides a form of relief, if not a belly laugh. The example of Gauss' solution generates a smile, as if to say, "My goodness. What an elegant approach. I appreciate it, but I never would have come up with it."

Even though it may be viewed as an act of genius (given the age and knowledge of Gauss at the time), it is something that can be appreciated by those who were about to attempt the problem in a more plodding way.

Still, it is interesting to wonder why such genius frequently emerges at an early age in mathematics while not so readily in fields that we normally conceive of as the humanities? Though we should be cautious about over-generalizing from this instance of talented behavior, the situation does have important components that may contribute to seeking alternatives to our oversubscribing problem solving as the end all and be all in mathematics.

For one thing the "machinery" needed for Gauss to exhibit his brilliant insight is relatively meager. That is, it was not necessary for him to know a great deal in order to put his ideas together in a clever and novel way. There was very little in the way of technical and formal understandings that were prerequisite to solving the problem. Though it was clever indeed, it was an unnoticed insight that was accessible to almost anyone who had experienced the adding of numbers. Though Gauss had surely added numbers before, as had all of his classmates, he most likely had thought of numbers and of operations on them differently. He most likely had viewed calculation problems not merely as tasks to be performed but as an invitation to see their dynamic relationships.

A second feature of the mathematical talent Gauss demonstrated is that the context within which he perceived the problem was essentially local and self-contained. It was local in the sense that it was not necessary to know where the problem came from, why it was given, why it was worth thinking about and what else flowed from it. It was self-contained in that it was perceived as a kind of experience that was not closely related to other experiences.

Mathematical genius is an intriguing concept in that it need not necessarily be focused so sharply on underlying structure or fundamental properties of a system as was the case with Gauss's discovery. In fact, it can flourish in an environment of *particularity* and the uniqueness of context as well. There is a wonderful story told by the British mathematician David Hardy (1877–1947), about Ramanujan, an Indian mathematician who made groundbreaking but untutored observations, frequently using unconventional symbolism. Newman (1956) quotes Hardy as saying about Ramanujan:

I remember once going to see him when he was lying ill at Putney. I had ridden in taxi-cab number 1729, and remarked that the number seemed to me rather dull one, and that I hoped it was not an unfavorable omen. "No," he replied, "it is a very interesting number; it is the smallest number expressible as a sum of two cubes in two different ways." (p. 375)

It is hard to know what was going on in Ramanujan's thinking that enabled him to come up with this observation. It is surely possible that he solved no perplexing problem at all in arriving at the observation that 1729 had a very unusual property. He may have "merely" recalled an observation he had made in his many previous calculations. It is possible that a lot of his thinking is of a sort that is inaccessible and virtually impossible for outsiders to discover—much as the mythology goes with regard to *idiot savants*. It is just as likely, however, that he was so immersed in thinking about numbers and about their relationships in an untutored way that he had created an enormous number of connections that enabled him to capture any particular number in his messy and inefficient but aesthetically appealing web.

Regardless of the manner in which the two concepts of genius appear, they do share features of isolation and simplicity, and in addition they do *not* require a sort of maturity (and frequently personal angst) that we associate with literature, philosophy, theology, and history. It is of course a far cry from the depiction of genius to selecting those qualities as the essential definition of a field.

Seeking incongruity in a larger domain than problem solving, however, does open up the possibility that we might find some interesting connections between mathematics and the (other) humanities, and that we might indeed expand the concept of genius beyond the nearly exclusive domain of problem solving. Our investigation of the incongruity in Process And Product—(2) above—opens a domain within which a new sort of talent might be exhibited. There we observed that Euclid's most powerful tool created to prove that there are an infinite number of primes, was one that was super non-prime.

Paradoxes are perhaps the sharpest example of mathematical incongruity. It is not that we seek out new and interesting ground to explore (as in juxtaposing process and product, in the case of Euclid's proof of the infinitude of primes; or in comparing mathematical and real world values, as in the case of Schnirelman's treasured find). Rather, in the case of paradoxes, the object of the paradox itself calls out to us since competing interpretations are so much at odds with each other.

Perhaps the most intriguing of all the categories of this section is that of self-referentiality. It surely beckons us to explore the depths of fields within the humanities that are normally thought to be inoculated against mathematical thinking. In particular it has the potential to enable us to explore aspects of self that do not normally reside within mathematics. But that's what humor has the capability to do as well.

In the section on punch lines, we will be discussing other ways of depicting mathematical thinking that have the potential to widen the concept of mathematical talent, to have the potential to better understand "self" as part of the journey.

PUNCH LINES

In humor, punch lines are thought of as the ending...especially with regard to jokes. They tie up neatly what preceded them. Usually the punch line provides closure. When it goes well, we are left with a feeling of having completed an experience. After hearing (or seeing, or perhaps even touching, tasting and smelling) it, we can look forward to encountering another new round that may be totally unconnected with the preceding one. The scope of a punch line is therefore limited.

What is like a punch line in mathematics? What should I take as the punch line of this essay? It might be best to leave answers to those questions themselves (returning to self-referentiality again) as punch lines for the reader to determine.

I am compelled to say a bit more, however, just so that I can view what usually provides closure for an experience as an invitation for a bit of "open-sure." In doing so, we could investigate the metaphorical punch line with regard to the large terrain of mathematics, or with regard to a particular branch of mathematics, or with regard to a subtopic within a branch, or with regard to a particular problem. We could even take all of these categories and transcend the concept of problem solving and proof.

My own teaching and writing has profited from the search for a range of metaphorical punch lines of the sort we just mentioned. If we think of punch lines as what we take away from an experience, then we might invite our students to think of how their mathematical explorations shed light on who they are as people. What sorts of punch lines do they seek? What sort of satisfaction do they derive from having acquired a punch line? In the case of jokes, most of us try to anticipate the punch line. Where do each of us situate the search for a punch line? At the end of a story? A collection of stories? Are we out for more than a punch line in itself, and perhaps try to use other people's punch lines as a way of understanding what intrigues them?

If as educators we knew that we were planning not only to ask those questions but to do so on a recurring basis, then it would affect what we teach and how we interact with subject matter and with our students. We would see ourselves differently.

Seeking forms of humor in mathematics has the potential to enable us to see deeper aspects of thinking and of feeling than we might ever otherwise imagined. Students who summarily depict mathematics as "easy" or "hard" for example lose sight of the interwoven fabric of mind that accompanies essentially every concept, as we discussed in most of the examples involving prime numbers. Thus, there is a significant difference between *understanding* a proof and punctuating the proof by noticing along the way intriguing building blocks that may have a life of their own. There is a difference more generally between *understanding* a proof and seeing it as an *aesthetic* object.

Furthermore, "coming up" with an idea is frequently powerful even though there may be no resolution or in fact no issue to resolve.

This, of course, raises the question of what objects entertain us in mathematical thinking/experiencing. Are they conjectures, proofs, questions, situations, items to behold, or mirrors of our minds?

Even when we are focused on conjectures, we can hold the conjecture in a abeyance for a while. At first when we explored Goldbach's conjecture, we glossed over the mundane observation that the sum of any two primes must yield an even number. But we then caught ourselves enmeshed in the mundane long enough to ask how it might be helpful. That happened when we used it to conjecture how Goldbach came up with his conjecture.

Goldbach's imagined act of inverting the mundane observation is a most powerful punch line. This is so even before the problem itself ("Can every even number be expressed as the sum of two primes?") is posed explicitly. It is a punch line in the sense that it opens up the potential for first putting a story to rest and then rejuvenating it. Up to that point, I had previously seen as trivial and not leading anywhere that the sum of two primes is an even number.

As we have seen, inverting some aspect of a concept in one way or another is a frequent bed-fellow of humor. It is helpful to view many long-standing problems in the history of mathematics from the perspective of punch lines.

In an important sense, what we have come to accept as a major portion of our school mathematics curriculum—extensions of number systems—as commonplace rather than as a long history of deleting and replacing punch lines. We sometimes forget that the very names we to give extended systems ("negative" numbers, "irrational" numbers, "complex" numbers, "imaginary" numbers) is testimony to the fact that these systems came about through years of labor pains, pains that were only relieved by coming to grips with systems that seemed anti-intuitive if not downright incongruous.

Take, for example, the early established and well entrenched intuition that a smaller number divided by a larger number cannot have the same value as a larger number divided by a smaller one. Now in an effort to "preserve" another well accepted property—that of "cross multiplication"—we find ways of persuading students that $(-1)/(1)$ and $(1)/(-1)$ are the same!

What is the logic that establishes the equality, and what is the price we must pay for reifying it? What is interesting about the above example (and all such examples requiring that we relinquish prior intuition as we expand what was previously not permissible) is that there is nothing God-given about the way in which one decides to adjudicate conflicting intuitions. In fact, we could create a respectable system of negative numbers, which maintained the prior conception of the ratio of larger to smaller numbers, but once again some other price would have to be paid.

There is a critical issue that lurks beneath the surface here that re-appears every time we try to extend number systems (in the above case to fractions and or negative numbers) so as to enable us to solve heretofore unsolvable problems. Consider, for example, at a slightly more advanced level, the question of how we go about assigning meaning to numbers such as $\sqrt{-1}$. Early on in student's education we find ways of persuading them that $\sqrt{-1}$ has no meaning, since the equation $x^2 = -1$ has no solution. If x is either positive or negative, its square is positive, and therefore cannot equal -1.

As students mature, however, we find ways of persuading them that in fact what had no prior meaning acquires meaning if we merely "extend" our number system to imaginary numbers. Well, how do we justify that extension? The issue is frequently clouded by the creation of some scheme (e.g., ordered pairs in which the first element is real and the second an imaginary number) that establishes the system, but by-passes the most fundamental logical issue: It *appears* that the new system (imaginary numbers) has all that the old scheme had and more—the more being that what was previously meaningless now acquires meaning through what appears to be an act of naming.

In fact, we *do* create a scheme that has logical force but at a price—the price being that we must relinquish something that we previously held to be dear—the property of *order*. That is, we can give $\sqrt{-1}$ and its fellow travelers meaning but only if we agree that we can no longer relate these numbers to zero nor to each other with regard to the concept of "greater than" or "less than."

As much of our algebraic curriculum can be viewed from the perspective of changing the arithmetic punch lines, so can we view the extension of Euclidean to non-Euclidean geometry.

The long history of non-Euclidean geometry was essentially an attempt to show that the parallel postulate was less primitive than the others. Efforts were made to *derive* that postulate from the earlier ones of Euclid. One of the most interesting digressions in the long history of the field is due to the work of an Italian geometer, Girolamo Saccheri (1667–1733). He published a book entitled, *Euclides Ab Omne Naevo Vindicatus,* in which he attempted to "vindicate Euclid of every blemish" by finally coming up with a proof of the parallel postulate (See Saccheri & Halsted, 1920).

Moïse (1963) describes the high humor associated with Saccheri's argument. First of all, there turned out to be a major flaw in his logic (having assumed without realizing it that he was making use of properties of straight lines that were not warranted by the postulates alone), and in fact he had not established the postulate as a theorem at all. Second Moïse argues, "The final irony is that if Saccheri's enterprise had succeeded in the way he thought it had, no modern mathematician would have regarded his book as a vindication of Euclid. From a modern viewpoint, a proof of the parallel postulate would merely show that the postulate was redundant; and redundancy is not thought of as a virtue..." (p. 131).

In the mid-nineteenth century, the problem of the parallel postulate was finally solved. It was solved in a way that truly vindicated Euclid according to what were then modern canons of mathematical logic. It was shown that the parallel postulate was independent of the other postulates. That means that it was proven that the postulate *could not* be derived from the others. Thus began the history of systems in which forms of the following assumptions were stipulated in different renditions of geometry:

1. Given a line with a point not on the line, there are many lines parallel to the given line.
2. Given a line with a point not on the line, there are no lines parallel to a given line.

Why, however, was this problem of establishing the role of the parallel postulate so resistant for centuries? The answer in large part is that the problem being posed for investigation was flawed. That is, the question that was being investigated was, "How can we prove the parallel postulate from the other ones?" It is a question that had unwarranted assumptions built in that took a long time to appreciate.

Another way of looking at the evolution of Euclidean geometry in light of the parallel postulate is that at various stages in its development, people saw punch lines that were later perceived to be premature. When and how we punctuate the tale of any mathematical idea is part of how we relate the tale to our expectations we have.

What is at stake with much of our discussion in this essay is an issue that is delicate but has more potential to humanize the act of problem-solving than heuristic and meta-cognitive schools have ever imagined. That is, extension of number systems and geometric entities is not merely a logical act by any means. In any act of extension, we are forced to confront an issue that resides at the intersection of logic and aesthetics. In realizing limitations of the machinery that we have created up to the extension point, we must do more than forge ahead and "create" something that was nonexistent or meaningless beforehand. In fact we must be cautious about imposing all property(ies) from the old system within that of the new. That is, we have to excise punch lines that we previously enjoyed.

Using various aspects of humor (*punch lines* and *irony* as our last examples of possible humor-connections) provides an intriguing lens not only through which to view mathematics but to raise some interesting questions about "self" as well.

Questions like the following might be some interesting starting points:

1. An important ingredient of humor is a sense of playfulness. How do the examples in this essay suggest strategies that could be adopted more generally to exhibit a sense of playfulness? How do some of these strategies function in non-mathematical settings? To what extent do you find yourselves enlarged/threatened/by the use of some of these strategies in mathematical as well as mathematical situations?

2. Though there is no iron-clad set of rules for generating ideas, questions and problems, what generating strategies have you used or might you use in a subject/topic you have recently explored? To what extent are they focused primarily upon proof in attempting to seek a punch line?

3. In true self-referential spirit, how would you modify the above questions (1 and 2 above), in a playful/humorous spirit so as to make them more (or perhaps less) enticing/challenging/clear to explore?

4. One generating strategy that is particularly powerful in humor and in mathematical thinking is one that I have discussed at length elsewhere: "What-if-not?" It involves looking at a concept, an idea, a problem, a solution, and asking yourself as an initial step: "What If Not?" Essentially it requires "seeing" something not only for what it *is* but as capable of becoming something else. A powerful example of this strategy applied to several of the questions about prime numbers can be found in Brown (1978, 1991). There I "tweaked" the domain from the set of *natural* numbers to the set E, {1, 2, 4, 6, 8...}, and compared conjectures about generating formulas, distribution of primes, Goldbach's conjecture and other related issues with those in E. The results were sometimes quite startling. Some unsolved problems in one domain relinquished their powerful hold in the other. Such tweaking is sometimes quite easy. At other times it is very difficult (as in the discussion of non-Euclidean geometry). Look back at some of the mathematical examples of your career and try to locate such thinking. See how it has functioned in aspects of your non-mathematical thinking (see Brown, S. I., 2001; Brown, S. I., & Walter, M. I., 1983, 1990, 2005).

5. What are some additional aspects of humor that we have not investigated? that you would be inclined to investigate?

6. Ludwig Wittgenstein *once* said (supposedly without being facetious) that a serious and good philosophical work could be written consisting entirely of jokes
   - Do you believe that he *ever* said it?
   - If so, do you believe that he said it only *once?*
   - What do you think he meant by it, if indeed he said it?
   - Would it have been possible to create this particular essay so that it would conform with Wittgenstein's suggestion? The introduc-

tion to "Black" Humor may be an example of something headed in that direction. Also, there was some concept associated with humor used in this essay that is an important ingredient of humor: that of *word play*. It was used here (without fuss) several times (as in referring to a "steal" platter, or discussing the perimeter of a piece of "Almond bread.") Can you find others (some are hidden)? Are there mathematical counter parts to"word play?"

7. Skim through the impressive array of books and essays (other than this one) associated in some way with Paul Ernest. Find one in which the role of humor in mathematics (education) is implied, or might be imposed or might suggest further exploration.

8. We discussed the issue of mathematical "talent" or genius emerging at an early age. The first example in the section on incongruity—was the famous example of Gauss. There we spoke of simplicity, clarity, isolation and an unexpected resolution. Of course, there the focus was upon problem solving in a narrow sense. We mentioned that the attributes that define this activity are not generally thought of as components of the humanities. It is possible that as we incorporate conventional mathematical thought with some of the ways of thinking we have described in this essay, that we will elaborate upon and even come up with new conceptions of mathematical talent— conceptions that are more consistent with ways we think in fields such as literature and philosophy. For a groundbreaking study that conceived of genius and talent according to these conventional categories, see Krutetskii (1976).

9. We have said very little in this final section about a theme that explicitly dominated the most introspection in our discussion of incongruity as a lynch pin to connect mathematics and humor: that of self-referentiality. It has, however been an undertow for most of this essay. I have, for example, tried to exhibit in my writing itself some of thee elements I have been drawing upon and analyzing. Without always doing so intentionally, is it possible that while I have not gone very far in writing a serious work that makes heavy use of jokes, could it be that Wittgensein would see my use of self-referentiality as an equivalent of joke-telling?

As a penultimate comment on opensure, it is worth noting that there is a critical underpinning for much of the above explorations that draws connections between mathematics and humor. That is, these questions for the most part challenge the assumption that mathematical thought is most valued when it is capable of being polished or finished. While it is the case that students understand that the curriculum "grows" and there is always more to learn, there is the general belief that we accomplish growth by moving ahead to new already explored territory. What we have been suggesting in essentially all of the examples and categories in this essay. is that any aspect of thought can be explored further if we see it not as being fixed, but in a state of transition.

Drawn by most definitions of the concept of "problem" itself, many educators assume that the category makes no sense unless problem solving is tied (perhaps loosely) to problem solving.

Much of what I have hinted at here and written about elsewhere deals with alternatives to a narrow view of mathematics as problem solving. In particular I have over the years discussed schemes that incorporate problem posing (sometimes loosely connected with problem solving and sometimes severing the connection altogether) as well. As a more extreme extension I have wondered what sorts of things one can do with a problem that depicts the category itself as being in a state of transition. At an extreme end, we can find ways of transforming a problem into something without a demand (to be solved).[13]

How will the concept of punch line enable us to extend this transformation further? What comes next after the concept of demand is softened as an implicit or explicit attribute of the concept of problem itself? There goes a final salute to self-referentiality.

I end this essay with a punch line of my own. It is a comment made by a student of mine at the end of a course and is one that I do not fully understand, but that I believe is worth taking seriously (which means—at very least—with humor). I had taught a mathematics course with a philosophical orientation to experienced teachers interested in the relation of mathematics to education. The course—described at the end of my 2007 Kpelle Tribe article—was focused on the Talmud as a format for exploring a number of mathematical themes. Though humor was not the actual subject matter of the course, I did make use of some of the strategies described in this essay. Much of what was done was not intentionally included, but I suppose that many of the ways of thinking and operating have become part of my being. The comments, however, say less about me in particular and more about the potential of humor to create new and more relaxed ways of experiencing education. For what it is worth, consider it to be a punch line for me, and a beginning for readers who would like to wonder further about humor and its role in mathematics and education.

This is truly a sad time. Yet still a time to rejoice. The semester is over. Yet so much has to be done. I am trying to think back to what I have immediately learned, and accepting that in the future, I will inevitably see more. One of the most peculiar things I have begun to develop, and which I credit this class. Is a better sense of humor and love of life in general. I had mentioned to Dr. Brown about laughing until crying and bodily convulsions set in. Some of the critical analysis skills I have picked up in this course have opened up these venues of humor. This course is titled for mathematics education, and I feel I have learned a lot of ways to improve upon my mathematics teaching. But I feel that most of what I learned was how to be a better person.... I feel that what this course has done is keep the playful spirit of the child in our education, and reminded us to keep it in our classrooms and our lives.

NOTES

1. See Brown (2007). Actually I had begun to write about humor and mathematics much earlier. Earlier. See the book review of John Allen Paulos' *Mathematics and Humor* by Stephen and Jordan Brown (1985). I wrote that review a quarter of a century ago together with my son, Jordan, who at the time was the first student in the US to establish the study of humor as an independent major.
2. See Brown (2001) for more elaborate discussion, p. 166–167. After reading about some of these theories, the reader may wish to examine the two jokes of Max Black to see what theory(ies) they exhibit. In addition to the authors whose work will be mentioned briefly in this section, see Cohen (1999) and Morreall (1987) for further discussion of various philosophical theories and dimensions of humor. Cohen includes an analysis of humor's moral dimensions.
3. The playful "an" in the parenthetical remark is included since it would be possible to view this entire problem from a geometric rather than algebraic point of view. It is worth wondering if the "essence" of the problem might be different depending upon which mathematical system we use. For example, a geometrical model would probably suggest quite different forms of generalization than an arithmetic one (see Brown, S. I. [1973] for an elaboration of this issue).
4. Actually, in 1937, Vinogradov improved a bit on Schnirelman's first crack. The results, however, are far from straightforward. That is, he showed that for any "sufficiently large" even number, the sum of at most four primes is necessary. Kramer (1970) discusses the context in which this proof was produced.
5. A variation of this problem and further discussion can be found in Brown (2001), p. 127–8.
6. With the growth of interest in intelligence design, there have been a number of recent books that take a sympathetic stance with regard to Sharon's query, and provide sympathetic and reasonable non- theistic responses. See Paulos' most recent book *Irreligion: A Mathematician Explains why the Arguments for God Just Don't Add Up* (2008). See also Barnes (2008), Dawkins (2006) and Hitchens (2007).
7. Actually, though I could not have done so at the time, if Sharon had asked the question about ten years later, it recently occurred to me that I might have asked her the following question: "If you place zero as a beginning point on a number line and then place all positive rational numbers in order on it from that starting point, what would be the first rational number you would hit when you move off the zero point?" What is enticing about this question

is, of course, that there is no "first" since you can always locate a smaller number between zero and it. What this does is to begin the conversation not only about countability, but about reasonable questions. On the surface, it seems that we should be able to find a first as we travel. That accords with our intuition about what happens when you walk along a path. Finding out that you cannot find a first, suggests that not all reasonable sounding questions have an answer. The fact that there is an assumption built into the question that is unwarranted, opens up all sorts of terrain about Sharon's original question about God.

8. Actually it is even reflected in the post colon title of his earlier book, *Gödel, Escher, Bach: An eternal golden braid* (1979).

9. Here is a brief popular summary of Russell's finding. He argued that there are essentially two types of sets. Specifically, some sets are members of themselves and some are not. For example, the set consisting of all ideas in the universe is itself an idea. Such a set is therefore a member of itself. The set consisting of all those sets that are meaningful is itself meaningful. This set also is therefore a member of itself. The set consisting of all irrational numbers is not itself an irrational number, however, and is therefore not a member of itself. Likewise the set consisting of all sets with less than five members has more than five members (in fact an infinite number) and therefore is also not a member of itself.

Now the genius of Russell's approach is that he created a new and curious set: the set of all sets that are not members of themselves. Let us call that set R (for Russell). He asked of this set an innocent sounding self-referential question: Is set R a member of itself?

If you think about that question in the privacy of your own boudoir, and if the light is shining at the right angle, and if you are in an upbeat mood, you may come to something like the following conclusion:

> If R is a member of itself, then since R consists only of those sets that are not members of themselves, R cannot be a member of itself. On the other hand, if R is not a member of itself, then R must be a member of itself since the set R captures only those sets that are not members of themselves.

Below is a concrete application of his set R:

> There is a town in which the barber shaves all those people and only those people who do not shave themselves. Now, who shaves the barber? If he shaves himself, then he cannot do so since he is to shave only those people who do not shave themselves. If he does not shave himself, then he shaves himself since he is to shave those people who do not shave themselves.

With this exposé, mathematicians and philosophers of mathematics could no longer feel secure in selecting sets as building blocks for the rest of the discipline. Russell created a concept called the theory of types in which he restricted the kinds of sets that it was possible to speak without fear of contradiction of the sort we discussed above.

10. A relatively non technical accounts of his proof and of the power of self-referentiality—the driving force in his argument—may be of interest. See Nagel and Newman (1958). See Dawson (1999) for a discussion of Gödel's genius and his psychological make-up.

11. Below is a standard proof by mathematical induction:

(i): P_1 is true. That is, if we have the trivial case of just the first term with which to contend, then,

$$1 = [1 \cdot (1 + 1)]/2.$$

(ii): Assume P_n.
Then we need to show that this implies P_{n+1}, as follows:
We would then examine:
$(1 + 2 + 3 + 4 + \ldots + n) + (n + 1)$.

Assuming P_n, we can re-express the above as
$[n\,(n + 1)]/2 + (n + 1)$.

By the distributive property we have:
$(n + 1) \cdot [(n/2 + 1)/]$

This is then simplified as:
$(n + 1) \cdot [(n + 2)/2]$,
which is of the form P_{n+1} since it is expressible as:
$[(n + 1) \cdot [(n + 2)]/2$.
Thus, the assumption that P_n is true does in fact lead us to conclude P_{n+1}.

12. Those who are persuaded that the scheme of mathematical induction is non-problematic might wonder about why it is necessary to establish P_1 since the second condition appears to be strong enough to do the job. I suggest the following conjecture about the sum of the integers in order to indicate that P_1 is necessary and also to suggest that the entire enterprise is more interesting than appears after one has been "acclimated" to its use:

$$1 + 2 + 3 + 4 + \ldots + n = [n \cdot (n + 1)]/2 + 17.$$

Now using the same kind of scheme for the second condition of mathematical induction we used to prove the valid equality, see what happens when you try to prove the above. The results will be surprising.

13. See my anecdote about the unwarranted power of viewing problems primarily as objects that have an inherent demand to be solved in Brown (2001), pp. 12–13.

REFERENCES

Barnes, J. (2008). *Nothing to be frightened of.* New York & London: Alfred A. Knopf.
Brown, S. I. (1969). Multiplication, addition, and duality. *The Mathematics Teacher* 59(6), 543–550, 591.

Brown, S. I. (1973). Mathematics and humanistic themes: *Sum* considerations. *Educational Theory, 23*(3), 191–214.

Brown, S. I. (1978, 1991). *Some "prime" comparisons.* Reston, VA: National Council of Teachers of Mathematics.

Brown, S. I. (2001). *Reconstructing school mathematics: Problems with problems and the real world.* NY: Peter Lang.

Brown, S. I. (2007). Transcending the Kpelle nightmare: Personal evolution and excavations. *Philosophy of Mathematics Education Journal 22,* 26. http://people. exeter.ac.uk/PErnest/pome22/index.htm

Brown, S. I., & Brown, J. D. (1985). *Mathematics and humor.* Book Review of John Allen Paulos (1980, University of Chicago Press). *Thinking: The Journal of Philosophy for Children 6*(1), 52–56.

Brown, S. I., & Walter, M. I. (1983). *The art of problem posing.* (with Marion Walter). Philadephia: The Franklin Institute Press.

Cohen, T. (1999). *Jokes: Philosophical thoughts on joking matters.* University of Chicago Press.

Dawkins, R. (2006), *The God delusion.* New York: Houghton Mifflin.

Dawson, J. W. (1999). Gödel and the limits of logic. *Scientific American, 280*(6), 76–81.

Federman, R. (1995), *Smiles on Washington Square.* Los Angeles: Sun & Moon Press.

Freud, S. (1960). *Jokes and their relation to the unconscious.* (originally published in 1905). NY: Norton.

Gleick, J. (1987). *Chaos:Making a new science.* Fairfield, PA: Penguin Books.

Greenhut, R. (Producer), & Allen, W. (Director), (1985). *The purple rose of Cairo* [Film]. Santa Monica, CA: Orion Pictures.

Heath, T. L. (1956). *The thirteen books of Euclid's elements.* New York: Dover.

Hitchens, C. (2007). *God is not great.* New York: Hatchette Book Group.

Hobbes, T. (1914). *Leviathan.* (originally published in 1651). London: Dent.

Hofstadter, D. R. (1979). *Gödel, Escher, Bach: An eternal golden braid.* New York: Basic Books.

Hofstadter, D. R. (2007). *I am a strange loop.* NY: Basic Books.

Kramer, E. E. (1970). *The nature and growth of modern mathematics.* New York: Hawthorne Books.

Krutetskii, V. A. (1976). *The psychology of mathematical abilities in school children.* In J. Kilpatrick & I. Wirszup (Eds.). *Soviet studies in the psychology of learning and teaching mathematics.* University of Chicago Press

Mandelbrot, B. (1999). A multifractal walk down Wall Street. *Scientific American, 280*(2), 70–74.

Mills, M. H. (1947). A prime-representing function. *American Mathematical Society Bulletin,* 53, 604.

Moïse, E. E. (1963). *Elementary geometry from an advanced standpoint.* Reading, MA: Addison-Wesley.

Morreall, J. (Ed.). (1987). *The philosophy of laughter and humor.* Albany: State University of New York Press.

Nagel, J., & Newman, J. (1958). *Gödel's proof.* New York University Press.

Newman, J. (1956). Srinivasa Ramanujan. In J. Newman (Ed.), *The world of mathematics.* (pp. 368–376). New York: Simon & Schuster.

Paulos, J. A. (1980). *Mathematics and humor.* University of Chicago Press.

Paulos, J. A. (2008). *Irreligion: A mathematician explains why the arguments for God just don't add up.* New York: Macmillan (Hill & Wang).

Saccheri, G. (author), & Halsted, B. (Ed.) (1920), Girolamo *Saccheri's Euclides Vindicatus.* Kila, MT: Kessinger.

CHAPTER 9

THE HUMAN CONDITION, MATHEMATICS, AND MATHEMATICS EDUCATION

Ubiratan D'Ambrosio
Brazil

*For Paul Ernest, as an homage on
his 65th birthday, which marks
a professional life dedicated
to improving the human condition.*

INTRODUCTION

This paper deals with humanness in the broad sense and how I see mathematics intrinsic to it. It is dedicated to Paul Ernest, who, as a philosopher of mathematics, recognized specificities of Mathematics Education and contributed, in a pioneering way, to the establishing the field of philosophy of mathematics education. I, particularly, owe much to Paul Ernest for his early receptivity of my proposals for ethnomathematics.

The possibility of final extinction of civilization in Earth is real. Not only through war. We are witnessing an environmental crisis, disruption of the economic system, institutional erosion, mounting social crises in just about

*Relatively and Philosophically E*ᵘ*rnest*, pages 127–145
Copyright © 2009 by Information Age Publishing

every country and, above all, the recurring threat of war. A scenario similar to the disruption of the Roman Empire is before us, with the aggravation that the means of disruption are, nowadays, practically impossible to control.

It is clear that mathematics is well integrated into the technological, industrial, military, economic and political systems and that mathematics has been relying on these systems for the material bases of its continuing progress. It is important to question the role of mathematics and mathematics education in arriving to the perverse behavior of mankind.

The denial and exclusion of the cultures of the periphery, so common in the colonial process, still prevails in modern society. The denial of knowledge that affects populations is of the same nature as the denial of knowledge to individuals, particularly children. Large sectors of the population do not have access to full citizenship. Some do not have access to the basic needs for survival. This is the situation in most of the world and occurs, even in the most developed and richest nations. To propose directions to counteract ingrained practices is the major challenge of educators, particularly mathematics educators. A new world order is urgently needed. What may be the role of mathematics in a new world order? I propose to look into the philosophy of mathematics and mathematics education with a broader view, accepting the fact that as mathematicians and mathematics educators we have to reflect about our personal role in reversing the threats to civilization.

THE RESPONSIBILITY OF MATHEMATICIANS AND MATHEMATICS EDUCATORS

Issues affecting society nowadays, such as national security, personal security, economics, social and environmental disruption, relations among nations, relations among social classes, people's welfare, the preservation of natural and cultural resources, and many others can be synthesized as Peace in its several dimensions: Inner Peace, Social Peace, Environmental Peace and Military Peace. These four dimensions of Peace are intimately related.

It is widely recognized that all these issues are universal, and it is common to blame, not without cause, the technological, industrial, military, economic and political complexes as responsible for the growing crises threatening humanity. Survival with dignity is the most universal problem facing mankind.

Mathematics, mathematicians and mathematics educators are deeply involved with all the issues affecting society nowadays. But we learn, through History, that the technological, industrial, military, economic and political complexes have developed thanks to mathematical instruments. And also that mathematics has been relying on these complexes for the material bases for its continuing progress. It is also widely recognized that mathematics is the most universal mode of thought. Although this is an Eurocentric vision, since what we are calling mathematics is the science which formally originated in the Mediterranean Basin about 3,000 years ago and spread to the entire world, it is the imprint of what we consider modern civilization.

It is sure that, as mathematicians and Mathematics Educator, we are concerned with the advancement of our discipline. But it is also sure that, as human beings, we are equally concerned with survival with dignity.

As a Mathematicians and Mathematics Educator I accept, as priority, the pursuit of a civilization with dignity for all, in which inequity, arrogance and bigotry have no place. This means, to reject violence and to achieve a world in peace.[1]

The Charter for a World without Violence, produced by recipients of the Nobel Peace Prize, states:

> We are convinced that adherence to the values of nonviolence will usher in a more peaceful, civilized world order in which more effective and fair governance, respectful of human dignity and the sanctity of life itself, may become a reality....

> In implementing the principles of this Charter we call upon all to work together towards a just, killing-free world in which everyone has the right not to be killed and responsibility not to kill others.

> To address all forms of violence we encourage scientific research in the fields of human interaction and dialogue, and we invite participation from the academic, scientific and religious communities to aid us in the transition to nonviolent, and non-killing societies.[2]

Scientists have been active leading the call for a stop in the insanity of war. Most pungent is the appeal of Albert Einstein and Bertrand Russell in the Pugwash Manifest, 1955: "We appeal, as human beings, to human beings: remember your humanity, and forget the rest." The Pugwash Movement or Pugwash Conferences on Science and World Affairs, which was awarded Nobel Price for Peace in 1995, has the motto "Thinking in a new way."[3]

I have no doubt that every mathematician and mathematics educators agree with me. Their discourse supporting this appeal is, without any doubt, sincere. But to go beyond wishful thinking and inspiring discourses, some innovative action is need.

Once we move, as mathematicians and mathematics educators, toward innovative action, something like a barrier appears, asking for continuing to do, with no more than superficial, indeed cosmetic, innovation, what as ever been done. This may obfuscate our concern, since it reinforces that for researchers, the priority is to publish their research in the best journals and, for mathematics educators, the priority is to theorize and develop methods, software and materials which aim at helping teachers to better prepare their students to pass the variety of tests which are imposed on them. In both constituencies, mathematicians and mathematics educators, sameness prevails!

As I said above, Peace must be understood in its multiple dimensions:

- Inner peace
- Social peace
- Environmental peace
- Military peace

My research program is to understand the responsibility of mathematicians and mathematics educators in offering venues for Peace. The Program Ethnomathematics, which will be discussed later in the paper, is a response to this.

A research program, on mathematics, history, education and on the curriculum, which is an attempt to face the question of responsibility, begins with a reflection on the nature of mathematical behavior. How is mathematics created? How different is mathematical creativity from other forms of creativity?

To face these questions there is need of a complete and structured view of the role of mathematics in building up our civilization, hence a look into the history and geography of human behavior.

I mean History and Geography not only as a chronological narrative of events and list of places, focused in the narrow geographic limits of a few civilizations, which have been successful in a short span of time. The course of the history of mankind can not be separated from the natural history of the planet. Indeed, the history of civilization has developed in close and increasing interdependence with the natural history of the planet. It is not so important to claim that although the Egyptian, the Sumerian and other civilizations were ahead of the Greek, the contribution to build up general mathematical theories was indisputably Greek.[4] As it is irrelevant that the medieval scholars received Euclid through the Arabs, what is largely accepted. What is very relevant is the fact that Mathematics as it is recognized today in the academia, developed parallel to Western thought (philosophical, religious, political, economical, artistic and, indeed, every sector of culture). Both, Mathematics and Western thought are the result of the dynamics of cultural encounters which occurred in the Mediterranean Basin and later in Central Europe. It would lead to a redundant boredom to give examples justifying this assertion. Indeed, Mathematics and Western Civilization belong to each other.

About Education, I claim that its major goals are:

- To promote creativity, helping people to fulfill their potentials and raise to the highest of their capability, but being careful not to promote docile citizens. We do not want our students to become citizens who obey and accept rules and codes which violate human dignity.
- To promote citizenship transmitting values and showing rights and responsibilities in society, but being careful not to promote irresponsible creativity. We do not want our student to become bright scientists creating new weaponry and instruments of oppression and inequity.

The big challenge we face is the encounter of the old and the new. The old is present in the societal values, which were established in the past and are essential in the concept of citizenship. And the new is intrinsic to the promotion of creativity, which points to the future. Hence, I see education as the dynamics of cultural encounters of the old culture, detained by parents, teachers, professional masters, naturally conservative, and the new culture, dominated by curiosity, imagination, hopes, characteristic of daughters and sons, students, apprentices. The challenge of education is to deal with this dynamics, which promotes progress.

I define curriculum as the strategy of educational systems to pursue these goals and to overcome the challenge of the dynamics of the encounter of the old culture and the new culture.

Curriculum is, usually, organized in three strands: objectives, contents, and methods. This Cartesian organization implies accepting the social aims of education systems, then identifying contents that may help to reach the goals and developing methods to transmit those contents.

THE POLITICAL DIMENSION
OF MATHEMATICS EDUCATION

The discussion on the objectives of Mathematics Education or, in other words, on "Why teach mathematics?" is regarded as the political dimension of education, but very rarely we see mathematics content and methodology been examined with respect to this dimension.[5] Regrettably, some educators and mathematicians claim that content and methods in mathematics have nothing to do with politics.

I feel very disturbing the possibility of leaving to the future generation, to our own children, a world convulsed by wars. Because mathematics conveys the imprint of Western thought, it is naïve not to look for a possible role of mathematics in framing a state of mind that tolerates war. As argued above, our major responsibility, as mathematicians and mathematics educators, is to offer venues to peace.[6]

I see my role as an educator and my discipline, mathematics, as complementary instruments to fulfill commitments to mankind. To make good use of these instruments, I must master them, but I also need to have a critical view of their potentialities and of the risk involved in misusing them. This is my professional commitment.

It is difficult to deny that mathematics provides an important instrument for social analyses. Western civilization entirely relies on data control and management. "The world of the twenty-first century is a world awash in numbers."[7] Social critics will find it difficult to argue without an understanding of basic quantitative mathematics.

Since the emergence of modern science, enormous emphasis has been placed on the rational dimension of man. Recently, multiple intelligences, emotional intelligence, spiritual intelligence, and numerous approaches to cognition, including new developments in artificial intelligence, challenge this. In mathematics education, this challenge is seen in the exclusive emphasis given to skill and drilling, as defended in some circles of mathematicians and mathematics educators.

HUMANNESS AND THE ESSENCE OF MATHEMATICS

I see mathematics as a fundamental instrument, both intellectual and material, in human knowledge and behavior. How did it develop?

Current systems of knowledge give support and a character of normality to the prevailing social, economical and political order. Both the religions and the sciences have advanced in a process of dismantling systems of knowledge, reassembling them and creating new systems of knowledge with the undeniable purpose of giving a sense of normality to prevailing human individual and social behavior.

The fundamental problem in this capability is the relation between brain and mind. It is possible to know much about the human body, its anatomy and physiology, to know much about neurons and yet know nothing why we like or dislike, love or hate. This gives rise to the modern theories of consciousness, which claims to be the last frontier of scientific research.[8]

Through a sophisticated communication system and other organic specificities, man tries to probe beyond the span of one's existence, before birth and after death. Here we find the origins of myths, traditions, religions, cults, arts and sciences. Essentially, this is a search for explanations, for understanding, which go together with the search for predictions. One explains in order to anticipate. Thus builds up systems of explanations (beliefs) and of behavior (norms, precepts). These are the common grounds of religions and sciences, until nowadays.

The drive towards survival is intrinsic to life. But the incursion into the mysteries beyond birth and death, which are equivalent to the search for past and future, seem to be typical of the human species. This is transcendence. The symbiotic drives towards survival and transcendence constitute the essence of being human.

The analysis of this symbiotic drive is focused in three elements, the INDIVIDUAL, the OTHER(S), organized as a SOCIETY, and NATURE, plus the three relations between them. Metaphorically, complex life may be represented by a triangle, emphasizing that the six elements are in mutual solidarity. The image of a triangle to relate basic components of the model is very convenient. I owe the idea for this triangle (the PRIMORDIAL TRIANGLE) as well as the ENHANCED TRIANGLE and the HUMANNESS TRIANGLE, which will be introduced below, to a paper of Antti Eskola.[9] A mathematical triangle ceases to be by the removal of any of the six elements. The same occurs with the life of an individual. It terminates with the removal of any of the six elements. Life ceases by the suppression of any of the three vertices or the interruption of the relation between them. The following image of the PRIMORDIAL TRIANGLE is very convenient.

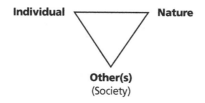

In species with developed neocortex, which we might call superior living species, the pulsion of survival, of the individual and of the species, and gregariousness, are genetically programmed. Reflexes, part of this programming, are usually identified as instinct.

The relations (sides) generate individual and social behavior. The triangle metaphor, meaning the indissolubility of the six elements, is resolved by the principles of physiology and ecology. Basically, the relation between individual and nature is responsible for nurturing, the relation of the individual and the other of opposite sex for mating and continuity of the species. Gregariousness is responsible for individuals organizing themselves in groups and herds, and hierarchy develops, most probably as an evolutionary strategy. The group, thus organized as society, relates to nature aiming at general equilibrium, following basic principles of ecology. Thus, the primordial triangle keeps its integrity. The rupture of each of the six elements eventually causes the extinction of a species.[10] Individual and social behaviors are actions taken "here" and "now."

Individuals of the human species, differently than other species with neocortexes, are provided, additionally, with will, that subordinates instinct.[11] Every individual has the ability to generalize and to decide actions that go beyond survival, thus transcending survival. Individuals acquire the sense of before/now/after and here/there. Individual and social behavior transcend here and now. Thanks to will, individuals develop preferences in nurture and in mating. They protect themselves and their kin and they plan ahead and provide. Physiological and ecological principles are not enough. Humans have to go beyond them and the relations (sides) and increment the primordial triangle by creating intermediacies. Between individual and nature, humans create instruments; language intermediates individual and the others; the relation between groups/society and nature is intermediated by production. In the process of recognizing the potential of these intermediacies, humans acquire an enlarged perception of nature. It becomes what is generally understood as REALITY, comprising natural, cultural and social environments. The primordial triangle becomes an ENHANCED TRIANGLE:

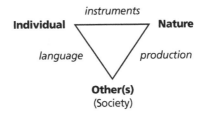

Instruments, both material and intellectual, which respond to will may be a conceptual image for technology. Instruments occur in dealing with a situation or need. An ad hoc solution develops, thanks to will, into technology. There are degrees of the response of technologies to what is expected of them.[12] The three intermediacies are, clearly, related. Instruments are shared through language and are decisive in the production system. The distinguishing feature of language is that it goes beyond mere communication and is responsible for the formation of new concepts. Language becomes essential in forming thought and determining personality features. It is in the roots of emotions, preferences and wants, which determines the enhanced relations of the individual and the other(s). Language is also essential in the definition and distribution of tasks, necessary for organizing systems of production. Thus, the intermediacies have also a form of solidarity which synthesizes what is called culture. Culture may be thus metaphorically expressed as a triangle, which I call the HUMANNESS TRIANGLE:

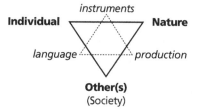

Human life is, thus, synthesized as the pursuit of the satisfaction of the pulsions of survival and transcendence. It is a mistake to claim, as many mathematicians do, that this refers to other forms of knowledge and that Mathematics has little to do with these pursues. A holistic view of History of Mathematics traces the origins of mathematics in pursuing the satisfaction of these two pulsions.

Engaging in survival, humans develop the means to work with the most immediate environment, which supplies air, water, food, necessary for nurturing, and with the other of opposite sex, necessary for procreation. These strategies, common to all superior living species, are absolutely necessary for the survival of the individuals and of the species. They generate modes of behavior and individual and collective knowledge, including communication, which is a complex of actions, utilizing bodily resources, aiming at influencing the action of others. In the species homo, behavior and knowledge include instruments, production and a sophisticated form of communication, which uses, as its means, language, as well as codes and symbols.

In the search of transcendence, the species homo develop the perception of past, present and future and their linkage, the explanations of and the creation of myths and mysteries to explain facts and phenomena encountered in their natural and imaginary environment. These are mentifacts incorporated to the individual memory and retrievable only by the individual who generated them. Material representations of the real, which we generally call artifacts, are organized as language, arts and techniques. Artifacts are observable and interpreted by others. In this process, codes and symbols are created. Shared mentifacts, through artifacts, have been called sociofacts by biologist Julian Huxley (1887–1975), who also introduced the words artifacts and mentifacts. The concept of culture for Huxley contemplates artifacts + mentifacts + sociofacts.

Explanations of the origins and the creation of myths and mysteries lead to the will to know the future [divinatory arts]. Examples of these arts are astrology, the oracles, logic, the I Ching, numerology and the sciences, in general, through which we may know what will happen—before it happens! The strategy of divinatory arts is deterministic.

Divinatory arts are based on mathematical concepts and ideas: observing, comparing, classifying, ordering, measuring, quantifying, inferring. Indeed these concepts and ideas are present in all the steps of the search for survival and transcendence.

As every form of knowledge, mathematical artifacts, in the form of practices and tools, and mentifacts, in the forms of aims or objectives, concepts and ideas, are first generated by individuals trying to cope and to deal with the natural and social environment, to resolve situations and problems, and to explain and understand facts and phenomena. These ad hoc artifacts and mentifacts are individually organized and are transmitted to other(s) and shared. They attain objectives, they serve, they are useful, they become methods, which are shared and acquired by the other(s), by society. They are part of the sociofacts of the group. How are they transmitted and shared? These are the basic questions when we ask for the origins of mathematics. Was the transmission and sharing through observation, mimicry? Eventually,using language. But, when? This is historically unknown. We have indications of the emergence of mathematical ideas thanks to artifacts, as it will be discussed later in this paper.

We have no idea when language was used in this socialization. Indeed, the origin of language was an academic "forbidden" theme about one hundred years ago. When language occurred, most probably systems of codes and symbols and specific words were created to design mathematical objects and ideas. This is major research subject for oral cultures. With the appearance of graphic registry, like cave drawings and bone carving, we have more elements to understand the development of mathematical concepts and ideas. The progress of mathematics through history, in different cultural environments, is a central issue to understand the nature of mathematics. In a recent book, Ladislav Kvasz discusses the historicity of linguistic tools as a major factor in the development of mathematics.[13] We may infer that, socially, this factor isolated mathematics from consideration of outsiders of the restricted circle of professional mathematicians, in a form of academic censorship. This kind of obstacle to critical views on mathematics and its social effects was already discussed above.

To share mathematics advances with the general population requires demystifying mathematics language. In an emblematic phrase, David Hilbert (1862–1943), probably the most eminent mathematician of the 20th century, said, in the major conference of the 2nd International Congress of Mathematicians:

> An old French mathematician said: "A mathematical theory is not to be considered complete until you have made it so clear that you can explain it to the first man whom you meet on the street."[14]

Demystifying mathematical language may open the way to a new form of mathematical education, with more space for critical analyses of mathematical development.

It is an undeniable right of every human being to share all the cultural and natural goods needed for material survival and intellectual enhancement. This is the essence of the United Nations' Universal Declaration of Human Rights.[15] The educational strand of this important profession on the rights of mankind is the World Declaration on Education for All, initially endorsed by 155 countries.[16] Of course, there are many difficulties in implementing United Nations resolutions and mechanisms. But as yet this is the best instrument available that may lead to a planetary civilization, with peace and dignity for all mankind.

THE ETHICAL DIMENSION OF MATHEMATICS EDUCATION.

It is not possible to relinquish our duty to cooperate, with respect and solidarity, with all the human beings who have the same rights for the preservation of good. The essence of the ethics of diversity is respect for, solidarity with, and cooperation with the other (the different). This leads to quality of life and dignity for all.

It is impossible to accept the exclusion of large sectors of the population of the world, both in developed and undeveloped nations. An explanation for this perverse concept of civilization asks for a deep reflection on colonialism. This is not to place blame on one or another, not an attempt to redo the past. Rather, to understand the past is a first step to move into the future. To accept inequity, arrogance, and bigotry is irrational and may lead to disaster. Mathematics has everything to do with this state of the world. A new world order is urgently needed. Our hopes for the future depend on learning—critically—the lessons of the past.

We have to look into history and epistemology with a broader view. The denial and exclusion of the cultures of the periphery, so common in the colonial process, still prevails in modern society. The denial of knowledge that affects populations is of the same nature as the denial of knowledge to individuals, particularly children. To propose directions to counteract ingrained practices is the major challenge of educators, particularly mathematics educators. Large sectors of the population do not have access to full citizenship. Some do not have access to the basic needs for survival. This is the situation in most of the world and occurs even in the most developed and richest nations.

THE PROGRAM ETHNOMATHEMATICS

A response to the responsibility of mathematicians and mathematics educators and a realization of this new concept of curriculum is the Program Ethnomathematics.

To build a civilization that rejects inequity, arrogance, and bigotry, education must give special attention to the redemption of peoples that have been, for a long time, subordinated and must give priority to the empowerment of the excluded sectors of societies.

The Program Ethnomathematics contributes to restoring cultural dignity and offers the intellectual tools for the exercise of citizenship. It enhances creativity, reinforces cultural self-respect, and offers a broad view of mankind. In everyday life, it is a system of knowledge that offers the possibility of a more favorable and harmonious relation between humans and between humans and nature.[17]

The Program Ethnomathematics offers the possibility of harmonious relations in human behavior and between humans and nature. It has; intrinsic to it; the ethics of diversity:

- Respect for the other (the different);
- Solidarity with the other;
- Cooperation with the other.

Let me elaborate on the genesis of this research program, which has obvious pedagogical implications.

An important question, frequently asked: is Ethnomathematics research or practice?

I see Ethnomathematics arising from research, and this is the reason for calling it the Program Ethnomathematics. But equally important, indeed what justifies this research, are the implications for curriculum innovation and development, teaching, teacher education, policy making and the effort to erase arrogance, inequity and bigotry in society.

For almost three decades, I have been formally involved with Pugwash Movement and the pursuit of peace (in all four dimensions: individual, social, environmental and military). A lecture of the History of Mankind makes it clear that Mathematics is central in all these dimensions. There is no need to elaborate on this.

An insight is gained by looking into non-Western civilizations. I base my research on established forms of knowledge (communications, languages, religions, arts, techniques, sciences, mathematics) and in a theory of knowledge and behavior which I call the "cycle of knowledge." This theoretical approach recognizes the cultural dynamics of the encounters, based on what I call the "basin metaphor." All this links to the historical and epistemological dimensions of the Program Ethnomathematics, which can bring new light into our understanding of how mathematical ideas are generated and how they evolved through the history of mankind. It is fundamental to recognize the contributions of other cultures and the importance of the dynamics of cultural encounters.

Culture, understood in its widest form, which includes art, history, languages, literature, medicine, music, philosophy, religion, science, technology, is characterized by shared knowledge systems, by compatible behavior and by acceptance of an assemblage of values. Research in ethnomathematics is, necessarily, transcultural and transdisciplinarian. The encounters of cultures are examined in its widest form, to permit exploration of more indirect interactions and influences, and to permit examination of subjects on a comparative basis.

Although academic mathematics developed in the Mediterranean Basin, expanded to Northern Europe and later to other parts of the World, it is difficult to deny that the codes and techniques which were developed, such as measuring, quantifying, inferring and the emergence of abstract thinking, as strategies to express and communicate the reflections on space, time, classifying, comparing, which are proper to the human species, are contextual. Clearly, in other regions of the World, other context give origin to different codes and techniques developed as strategies to express and communicate the reflections of a different spatial context, a different time perception, and different ways of classifying and comparing. These are, obviously, contextual.

At this moment, it is important to clarify that my view of ethnomathematics should not be confused with ethnic-mathematics, as it is mistakenly understood by many. This is the reason why I insist in using Program Ethnomathematics, which tries to understand and explain the various system of knowledge, such as mathematics, religion, culinary, dressing, football and several other practical and abstract manifestations of the human species in different contextual realities. Of course, the Program Ethnomathematics was initially inspired by recognizing ideas and ways of doing that reminds us of Western mathematics. What we call mathematics in the academia is a Western construct. Although dealing with space, time, classifying, comparing, which are is proper to the human species, the codes and techniques to express and communicate the reflections on these behaviors is undeniably contextual. I got an insight into this general approach while visiting other cultural environments, during my work in Africa, in practically all the countries of continental America and the Caribbean, and in some European environments. Later, I tried to understand the situation in Asia and Oceania, although with no field work. Thus, came my approach to Cultural Anthropology (curiously, my first book on Ethnomathematics was placed by the publishers in a collection of Anthropology).

To express these ideas, which I call a research program, I created a neologism, *ethno + mathema + tics*. This caused much criticism, because it does
not reflect the etymology of "mathematics." Indeed, "mathematics" is not
composed, it is a neologism, with Greek origin, introduced in the XIV century. It is not mathema + tics. The idea of organizing these reflections occurred to me while attending International Congress of Mathematicians
ICM 78, in Helsinki. In playing with Finnish dictionaries (to play with dictionaries is a favorite pastime), I was tempted to write alustapasivistykselitys
for the research program. Bizarre! So, I believed Ethnomathematics would
be more palatable.

I understand that there are immediate questions facing World societies and education, particularly mathematics education. As a mathematics
educator, I address these questions. Thus, the Program Ethnomathematics
links to the study of curriculum, and to my proposal for a modern trivium:
literacy, matheracy and technoracy.[18]

The pursuit of Peace, in all four dimensions mentioned above, is an
urgent need. Thus, the relation of the Program Ethnomathematics with
Peace, Ethics and Citizenship.

These lines of work in mathematics education link, naturally, to the pedagogical and social dimensions of the Program Ethnomathematics.

As I said above, it is important to insist that the Program Ethnomathematics is not ethnic mathematics, as some commentators interpret it. Of
course, one has to work with different cultural environments and, as an
ethnographer, try to describe mathematical ideas and practices of other
cultures. This is a style of doing ethnomathematics, which is absolutely necessary. These cultural environments include not only indigenous populations, but labour and artisan groups, communities in urban environment
and in the periphery, farms, professional groups. These groups develop
their own practices, have specific jargons and theorize on their ideas. This
is an important element for the development of the Program Ethnomathematics, as important as the cycle of knowledge and the recognition of the
cultural encounters.

Basically, investigation in ethnomathematics start with three basic questions:

1. How are ad hoc practices and solution of problems developed into
   methods?
2. How are methods developed into theories?
3. How are theories developed into scientific invention?

It is important to recognize the special role of technology in the human
species and the implications of this for science and mathematics. Thus, the
need of History, of Science, and Technology (and, of course, of Mathemat-

ics) to understand the role of technology as a consequence of science; but, also as an essential element for furthering scientific ideas and theories.[19]

Once recognized the role of technology in the development of mathematics, reflections about the future of mathematics propose important questions about the role of technology in mathematics education. Besides these more immediate concerns, there are long term concerns. Of course, they are related. Hence, the importance of linking with Future Studies. And, in particular, with Distance Education.

Reflections about the presence of technology in modern civilization leads, naturally, to question about the future of our species. Thus, the importance of the emergent fields of Primatology and Artificial Intelligence, which lead to a reflection about the future of the human species. Cybernetics and human consciousness lead, naturally, to reflections about fyborgs (a kind of "new" species, i.e., humans with enormous inbuilt technological dependence).[20] Our children will be fyborgs when, around 2025, they become decision makers and take charge of all societal affairs. Educating these future fyborgs calls, necessarily, for much broader concepts of learning and teaching. The role of mathematics in the future is undeniable. But, what kind of mathematics?

To understand how, historically, societies absorb innovation, is greatly aided by looking into fiction literature (from iconography to written fiction, music and cinema). It is important to understand the way material and intellectual innovation permeates the thinking and the myths, and the ways of knowing and doing of non-initiated people. In a sense, how new ideas vulgarise, understanding vulgarise as making abstruse theories and artefacts easier to understand in a popular way.

How communities deal with space and time, mainly to understand the sacralization of chronology and topology in history, is also central.

We have to look into the cultural dynamics of the encounter of generations (parents and teachers and youth). This encounter is dominated by mistrust and cooptation, reflected in testing and evaluation practices, which dominate our civilization. In mathematics education, this is particularly disastrous. Mathematics is, usually, seen by youth as uninteresting, obsolete and useless way. And they are right. Much of is in the traditional curriculum is uninteresting, obsolete and useless.

Resources to testing, is the main argument to justify current math contents. The claims of the importance of current math contents are fragile. Myths surround these claims.

It is important to understand children and youth behavior and their expectations. History gives us hints on how periods of great changes affect the relations between generations. Most interesting is the analysis of youth movements as a reaction to the same tensions in international relations, while the wounds of World War II and Viet Nam were still open. Of major importance is the 1968 revolt in Paris, which spread to many other cities.

Regrettably, education, in general, is dominated by a kind of "corporate" attitude, in the sense that there is more concern with the subjects taught than with the children. This is particularly true with Mathematics Education. There is more concern with attaining pre-decided goals of proficiency, which favours sameness and may lead to the promotion of docile citizens and irresponsible creativity. Tests are the best instruments to support this corporate aspect of education. Tests penalize creative and critical education, which leads to intimidation of the new and to the reproduction of this model of society.

THE CRITICS OF ETHNOMATHEMATICS

What to say to critics who dismiss ethnomathematics as political correctness gone too far?

It is difficult to deny that mathematics, as well as education in general, are the arms of a political and ideological posture. Ethnomathematics is no different. Yes, ethnomathematics is political correctness. If proposing a pedagogical practice which aims at eliminating truculence, arrogance, intolerance, discrimination, inequity, bigotry and hatred, is labeled as going too far, what to say?

Is it questionable to refer to truculence, arrogance, intolerance, discrimination, inequity, bigotry, hatred, when discussing mathematics? Then the question is not pedagogical, but historical. As I said above, the historiography of mathematics has been very conservative and biased. Of course, both pedagogical and historical issues are related (see D'Ambrosio, 1998b).

What to say to critics who charge that ethnomathematics does little to advance students' knowledge and understanding of mathematics?

It is clear that traditional teaching of mathematics is not satisfying. Testing and assessment is part of the traditional teaching. The alarming results of tests are the result of a very poor education, which is performed in the traditional methods and curricula. Ethnomathematics do not reach a substantial student population and do not have any effect in the bad results of testing. Measures to "tighten" traditional teaching, hoping to get better results in tests and assessment, are nothing less than disastrous. Countries which are model of traditional teaching and are proud of their systems, are the most vulnerable.

It is clear that reinforcing sameness is not the answer to vulnerability. Although history tells this, and as examples I mention the fall of the Roman empire, the collapse of the Third Reich, the fiasco of the Soviet interference in Afghanistan, the demolition of the Berlin Wall, and others. Sameness, like fundamentalism, lacks creativity to counter vulnerability. Tightening measures lead to worsening effects.

This is particularly true in education. Insisting in obsolete, uninteresting and useless mathematics education, will not avoid its rejection by the new generations. On the other hand, by focusing on individual dignity, recognizing the previous knowledge of the individual and of her/his culture [ethnomathematics], we can prepare the most fertile ground for building up new knowledge (mathematics).

It is an important step in education to recognize that all forms of knowledge, both ethnomathematics and mathematics as well, have limitations. So, it is natural to look for new communicative and analytic instruments. This is why history of mathematics and ethnomathematics should be together. Every advance in mathematics is related to overcoming difficulties in doing or explaining something. The advancement of knowledge and understanding of mathematics, once the ground is fertile, is a matter of motivation. Has much more to do with the overall goals and objectives of mathematics education. Why to deny ethnomathematics, which is clearly alive in the professions, in communities, in extant cultures and in cultural history? Who is afraid of it?

The reason to teach ethnomathematics of other cultures, for example, the mathematics of ancient Egypt, the mathematics of the Mayas, the mathematics of basket weavers and ceramists, of professional practitioners, as surgeons and architects, is not because they are important for children. It is because there is a deep educational reason for this. The importance is because to learn other ethnomathematics helps the individual to improve the understanding of its own. This is related to an apparent contradiction of proposing ethnomathematics when the mood of the world is globalization.[21]

Like languages, if we domain only one language, we are less equipped to succeed in the modern World than if we have some proficiency in other languages. And it is a known fact that knowing other languages is a positive factor in bettering the domain of one's own language.

The main reasons for ethnomathematics in the curriculum are:

1. To desmistify a form of knowledge [mathematics] as being final, permanent, absolute, unique. There is a current misperception in societies, very damaging, that those who perform well in mathematics are more intelligent, indeed "superior" to others. This erroneous impression given by traditional mathematics teaching is easily extrapolated to religious, ideological, political, racial creeds;

2. To illustrate intellectual achievement of various civilizations, cultures, peoples, professions, gender. Mathematics is absolutely integrated with Western civilization, which conquered and dominated the en-

tire world. The acceptance, forced or voluntary, of Western knowledge, behavior and values, can not be associated with ideas like "the winner is the best, the losers are to be discarded". More than any other form of knowledge, mathematics is identified with winners. This is true in history, in the professions, in everyday life, in families, in schools. The only possibility of building up a planetary civilization depends on restoring the dignity of the losers and, together, winners and losers, moving into the new. This requires respect for each other. Otherwise, the efforts will be from the losers to become winners, and from the winners to protect themselves from the losers, thus generating defensive confrontation.

Ethnomathematics practices in school favour respect for the other and solidarity and cooperation with the other. It is thus associated with the pursuit of PEACE. The main goal of Ethnomathematics is building up a civilization free of truculence, arrogance, intolerance, discrimination, inequity, bigotry and hatred.

NOTES

1. *Pugwash Conferences on Science and World Affairs.* http://www.pugwash.org/ (retrieved 03 March 2009).
2. Charter for a World Without Violence http://www.ipb.org/8thWorldSummit Nobel/Charter%20of%20World%20Without%20Violence.pdf (retrieved 03 March 2009).
3. The Russell-Einstein Manifesto, *Scientists in the Quest for Peace. A History of the Pugwash Conferences.* Ed. Joseph Rotblat, Cambridge: The MIT Press, 1972; pp. 137–140.
4. This is the main issue of the polemics about Afrocentrism. See Mary Lefkowitz: *Not Out of Africa. How Afrocentrism Became an Excuse to Teach Myth as History,* New York: Basic Books, 1996.
5. This was the theme of my conference in ICME 3: Overall Goals and Objectives of Mathematics Teaching, *Proceedings of the Third International Congress of Mathematics Education,* Karlsruhe, 1976, pp. 221–227.
6. Ubiratan D'Ambrosio: Mathematics and Peace: Our Responsibilities, *Zentralblatt für Didaktik der Mathematik/ZDM,* 30(3): 1998; pp. 67–73.
7. Lynn Arthur Steen, ed.: *Mathematics and Democracy: The Case for Quantitative Literacy.* Princeton, NJ: National Council on Education and the Disciplines, 2001; p. 1.
8. See the important oeuvre of Oliver Saks, particularly *An Anthropologist on Mars.* Theories of consciousness also give rise to several academic controversies. See for example the review by David Papineau of the book by David J. Chalmers: *The Conscious Mind. In search of a fundamental theory,* Oxford University Press,

1995, which was published in *The Times Literary Supplement,* June 21, 1996, pp. 3–4 as "A universe of zombies?"

9. Antti Eskola: Civilization as a Promise and as a Threat, *peaceletter* (Helsinki), 1/89, pp. 8–14.

10. It is worthwhile to look, for inspiring reflections, see the novel of emeritous paleontologist George Gaylord Simpson: *The Dechronization of Sam Magruder,* St Martin's Press, New York, 1995.

11. Will is a recurrent theme in philosophy, religion, neurosciences.

12. The degrees of response are well studies, through the concepts of whole technology and broken technology, by Fernando Flores Morador: *Broken Technologies. The Humanist as Engineer,* Lund University, 2009.

13. Ladislav Kvasz: *Patterns of Change. Linguistic Innovations in the Development of Classical Mathematics,* Birkhäuser, Basel, 2008.

14. David Hilbert: Mathematical Problems, *Bulletin of the American Mathematical Society.* July 1902, p. 438.

15. United Nations: *Universal Declaration of Human Rights,* 1948 http://www.un.org/Overview/rights.html (retrieved 03 March 2009).

16. UNESCO: *World Declaration on Education for All,* 1990, http://www.unesco.org/education/efa/ed_for_all/background/jomtien_declaration.shtml (retrieved 03 March 2009).

17. Ubiratan D'Ambrosio: Ethnomathematics and its First International Congress. *Zentralblatt für Didaktik der Mathematik, ZDM.* 31(2), 1999; pp. 50–53.

18. Ubiratan D'Ambrosio: Literacy, Materacy, and Technoracy: A Trivium for Today. *Mathematical Thinking and Learning,*1(2), 1999; pp. 131–153.

19. Ubiratan D'Ambrosio: Ethnomathematics and its Place in the History and Pedagogy of Mathematics, *For the Learning of Mathematics,* vol. 5, nº1, February 1985, pp. 44–48.

20. The term was first used by Alexander Chislenko (1959–2000) in 1995.

21. Ubiratan D'Ambrosio: The Program Ethnomathematics and the challenges of globalization, *CIRCUMSCRIBERE, International Journal for the History of Science*vol.1, 2006, pp. 74–82 http://www.pucsp.br/pos/cesima/circumscribere (retrieved 03 March 2009).

CHAPTER 10

ON FIELD(ING) KNOWLEDGE

Sharon Friesen and David W. Jardine
University of Calgary, Canada

"Education is suffering from narration-sickness," says Paulo Freire.
It speaks out of a story which was once full of enthusiasm, but now shows itself
incapable of a surprise ending. The nausea of narration-sickness comes from having
heard enough, of hearing many variations on a theme but no new theme.
—David G. Smith (1999, p. 135)

INTRODUCTION

The authors have worked together, now, for nearly twenty years. In many settings we've found our individual and collective way around classrooms from Kindergarten to Grade Twelve, through undergraduate classes in teacher preparation programs, up through Masters and Ph.D. level work. Sharon has been deeply involved with school assessments, provincial and cross-provincial curriculum development and implementation. Our work has been persistently (re)composed around an evolving question: How can mathematics be experienced in the classroom as a living field of knowledge? We think, for example, of some of the work that was coincident with our first connections with Professor Ernest—early, now-ancient-seeming mediations on the analogical and generative kinship fields of mathemati-

Relatively and Philosophically E'rnest, pages 147–172
Copyright © 2009 by Information Age Publishing

147

cal language (Jardine, 1990; for an expanded version, see Jardine, 1994) and early thoughts on the nature of time and its passing and how to enact a classroom practice of such matters out from under the spell of simply "managing" (Clifford & Friesen, 1993). In that 1990 paper entitled "On the Humility of Mathematical Language," the eco-analogical character of mathematical language showed itself to be a wide, complex field of kinships and interrelationships that are not reducible to a univocal, singular sense. It was Professor Ernest's generous response to this paper after its publication (a long, handwritten, red-ink missive sent in the post) that provided such encouragement for both of us to continue exploring this territory.

Over the intervening years we've explored folds of the same cloth: "family resemblances" (from Wittgenstein, 1968; see Jardine, 1992a, graciously re-published and introduced by Professor Ernest as Jardine, 2004), ancestries, fields, topographies, places, big ideas (Clifford & Friesen, 1993; Jardine, Friesen, & Clifford, 2003), generative/intergenerational knowledge (Friesen, Clifford, & Jardine, 1998), living disciplines, relatedness and interdependence (see Jardine 1995, Jardine & Friesen, 1997), ecopedagogy (Jardine, 1997, 2000), and the ontologies of mathematics that underwrite such fields of knowledge. We've become taken by the idea that mathematics is a living inheritance and must be treated as such to be properly understood:

> [Mathematics does not have] the character of an object that stands over and against us. We are no longer able to approach this like an object of knowledge, grasping, measuring and controlling. Rather than meeting us in our world, it is much more a world into which we ourselves are drawn. [It] possesses its own worldliness and, thus, the center of its own Being so long as it is not placed into the object-world of producing and marketing. The Being of this thing cannot be accessed by objectively measuring and estimating; rather, the totality of a lived context has entered into and is present in the thing. And we belong to it as well. Our orientation to it is always something like our orientation to an inheritance that this thing belongs to, be it from a stranger's life or from our own. (Gadamer, 1994, pp. 191–192)

The authors, together and separately (and often with—and henceforth, sadly, without our fellow traveler Patricia Clifford), we have been trying to trace a tale of mathematics centered around this undeniable phenomenological datum: being drawn into a world, a living field (see Clifford & Marinucci, 2008). This datum, familiar, we suggest, to those experienced in mathematics, sidesteps some of the degenerative educational traditions we have inherited that have drained the life out of what we have always experienced as a *living* discipline. We're interested in the nature of the "continuity of attention and devotion" (Berry, 1986, p. 33) these matters require and deserve.

VARIATIONS ON A THEME BUT NO NEW THEME

To begin a story, someone in some way must break a particular silence.
—Wiebe & Johnson, 1998, p. 4

In these long and varied travels, we have come to know something of narration sickness, and how a once enthusiastic tale of the ways of schools has become ever-increasingly, nauseatingly numbing. We have been intimately involved with hundreds of teachers and students and have witnessed, first hand, an old tale, which was once full of enthusiasm, still holding sway: a tale of fragmentation, breakdown, linearity, and literalism, coupled with regimes of surveillance, management and its requisite standardization of assessment, and all the consequent sicknesses. Students have become ill, dull, disinterested in the face of this tale. Teachers, too, have become ill. And what is taken to be 'mathematics' has itself fallen pallid and weak, infected with a industrial assembly-line story-line that has trumped its own living ways. Perhaps even more insidious is how the (often silent) dominance of this story-line allows for the assignation of blame for such ills on the sufferers themselves.

We've recently been searching back once again into the origins of this persistent and now pernicious story-line. It is a story-line now trumpeted often in schools as simply "the way things are," "the real world." Thus cast under the "sickness of literalism" (Hillman, 1989, p. 3), it is a story that has forgotten and fallen silent of its telling. All we can do is try to break this particular silence again.

We've come upon the figure of Fredrick Winslow Taylor (1856–1915).

In the late-19th and early 20th century industrial world of the East Coast United States, from work with Bethlehem Steel Company (circa 1899) and later with Henry Ford (on Taylorism and Fordism, see Kanigel, 2005, p. 498), Taylor instituted what was later to be called 'the efficiency movement' (Callahan, 1964). This movement arose out of Taylor's observations on the shop floors of various industries and his development of what he called time and motion studies. In order to make such industries more efficient and less wasteful of time, materials and energy, Taylor, with a clipboard and stopwatch in hand, moved to break down any particular task into its component parts and lay out ways in which the organization, management and sequencing of that task could be more efficiently organized. This required regimes of standardization, surveillance, sequencing, and many other now-familiar consequences. It lead to shop managers being able to assess any finished product and target any errors in production and precisely assess and locate the source of such errors. Its most pristine cultural image is that of Henry Ford's assembly line: each worker has placed in front of them an isolated, repeated task to be done with singular, standardized procedures and invariant materials. All tethers of this task to the object be assembled have been severed vis a vis the workers work. Such tethers and relationships are a management issue only. The task for workers is simply to learn by rote and repetition the efficient accomplishment of this one, isolated task.

This prospect of efficiency swept through all facets of then-contemporary life, from mayor's offices to hospitals to, of course, schools. Under the pressures of early 20th century immigration, coupled with burgeoning needs for workers in every-expanding industrial facilities, the promise of efficiency was irresistible: "educators needed little prompting" (Dufour & Eaker, 1998, p. 20). Given the burgeoning numbers of immigrant children into large East Coast American cities, and the equally burgeoning need for minimally educated workers in industry, schools had become overwhelmed early in the 20th century, and the prospect of a more manageable, efficient organization of schooling became irresistible.

By the time of the publication of Taylor's *The Principles of Scientific Management* in 1911, the clarion of efficiency had become a widespread public fad and fancy (see Gatto, 2006; Callahan 1964). The promise of "the one best way" (Kanigel, 2005) in Taylor's work and ideas had moved from the limits of industrial production to any form of organization whatsoever. Articles were published regularly in both scholarly journals and popular magazines touting the significance of this new phenomenon of "efficiency" and supporting its willy-nilly application to any and all social ills.

Specifically with regard to education, we have, for example, the words of Ellwood P. Cubberley, dean of the school of education at Stanford, from his book *Public School Administration*, originally published in 1916 (cited here from Callahan 1964, p. 97; see also Cubberley, 1922):

> Our schools are, in a sense, factories in which the raw products (children) are to be shaped and fashioned into products to meet the various demands of life. The specifications for manufacturing come from the demands of twentieth-century civilization, and it is the business of the school to build its pupils according to the specifications laid down. This demands good tools, specialized machinery, continuous measurement of production to see if it is according to specifications, [and] the elimination of waste in manufacture.

Parallel initiatives were accomplished in the area of language by Leonard Ayres in his *A Measuring Scale for Ability in Spelling* (1915) (see also his *Laggards In Our Schools* [1909], the full text of which is available on-line). In the former work, the tethers of sense and significance between words being learned by children were cut. Individual words were tested in isolation for the statistical frequency of their usage and the statistical frequency of errors made by those learning these words in isolation. Thus isolated, the only shape which the learning of such words took was a version of the now school grade by school grade assembly of this list. Spelling lists for various grades are still organized around this work. In fact, arguments about children's recent inability to spell as well as children in the early 20th century has been attributed to an increasing failure to stick with these lists and their implicit pedagogy (see Geraldine Rodgers's, 1984 text compiled in; Potter, 2004; and also Rodgers, 1983. Both of Rodgers' texts are calls to return to Ayres scales; the complete text of Ayres *Measuring Scale* is available on Potter's website: donpotter.net). "If the results show that present-day children are not achieving as Ayres' 1914–1915 children did, we should not conclude that it is [present-day] children who are defective. It is [present-day] teaching methods which are defective" (Rodgers, 1984; complied by Potter, 2004).

"Taylor's thinking so permeates the soil of modern life we no longer realize it's there. It has become, as Edward Eyre Hunt, an aide to future President Herbert Hoover, could grandly declaim in 1924, 'part of our moral inheritance'" (Kanigel, 2005, p. 7). This might provide a first clue regarding the sort of moral indignation we have run into when we attempt to interrupt this breakdown-efficiency inheritance and name it and its effects. Consider this passage from H. Martyn Hart, the Dean of St. John's Cathedral in Denver, from the September 1912 issue of the *Ladies' Home Journal*:

> The system [of schooling] . . . has indeed become a positive detriment and is producing a type of character which is not fit to meet virtuously the temptations and exigencies of modern life. The crime which stalks almost unblushingly through the land; the want of responsibility which defames our social honor; the appalling frequency of divorce: the utter lack of self-control; the abundant use of illicit means to gain political positions; are all traceable to its one great and crying defect—inefficiency. (cited in Callahan 1964, p. 52)

From Taylor and various commentaries on his work, we get broad strokes that are all too familiar in many school settings:

> In the type of management advocated by the writer, this complete standardization of all details and methods is not only desirable but absolutely indispensable as a preliminary to specifying the time in which each operation shall be done, and then insisting that it shall be done within the time allowed. Neglecting to take the time and trouble to thoroughly standardize all of such methods and details is one of the chief causes for setbacks and failure in introducing this system. (Taylor, 1903)

> "Every day, year in and year out, each man should ask himself over and over again, two questions," said Taylor in his standard lecture. "First, 'What is the name of the man I am now working for?' And having answered this definitely

then 'What does this man want me to do, right now?' Not, 'What ought I to do in the interests of the company I am working for?' Not, 'What are the duties of the position I am filling? Not, 'What did I agree to do when I came here?' Not, 'What should I do for my own best interest?' but plainly and simply, 'What does this man want me to do?'" (cited in Boyle, 2006)

From a June 4, 1906 lecture (cited in Kanigel, 2005, p. 169):

In our scheme we do not ask for the initiative of our men. We do no want any initiative. All we want of them is to obey the orders we give them, do what we say, and do it quick.

"His declared purpose was to take all control from the hands of the workman (whom he regularly compared to oxen or horses) and place it in those of management" (Kanigel, 2005, p. 19).

Students and teachers, then, are not required to be thoughtfully engaged in teaching and learning. Long, drawn out conversations or explorations are understood as simply *means* by which rote factual and procedural information is to be attained. Such conversations are understood, therefore, as, basically, an inefficient "a waste of time." Or, as many teachers have said to us, they would love to have such conversations in their classrooms, but this is "an exam year" and, in the work of the assembly-line of schools, "time is always running out" (1983, p. 76).

Thus schooling itself became subjected to a profound form of anti-intellectualism, where any talk of mulling over the ways of mathematical knowledge as a living field is seen as an opulent, unnecessary venture:

Suddenly school critics were everywhere. A major assault was mounted in two popular journals, *Saturday Evening Post* and *Ladies Home Journal*, with millions each in circulation, both read by leaders of the middle classes. The *Post* sounded the anti-intellectual theme this way:

Miltonized, Chaucerized, Vergilized, Shillered, physicked and chemicaled, the high school. . . . should be of no use in the world—particularly the business world.

Three heavy punches in succession came from *Ladies Home Journal*: "The case of Seventeen Million Children—Is Our Public-School System Providing an Utter Failure?" This declaration would seem difficult to top, but the second article did just that: "Is the Public School a Failure? It Is: The Most Momentous Failure in Our American Life Today." And a third, written by the principal of a New York City high school, went even further. Entitled "The Danger of Running a Fool Factory," it made this point: that education is "permeated with errors and hypocrisy," while the Dean of Columbia Teachers College, James E. Russell added that "If school cannot be made to drop its mental development obsession the whole system should be abolished." (Wrege & Greenwood, 1991)

"What [Taylor] really wanted working men to be [is] focused, uncomplicated and compliant" (Boyle, 2006). Even though this story has lost much of its telling character, in the setting of schools still bent on efficiency, it is teachers and students alike who must needs be thus uncomplicated and compliant. Such anti-intellectualism (see Callahan, 1964, p. 8) regarding schools and their purpose, such a determined orientation of schooling to the "world...of business" and its wants still haunts schooling, even as we find that such blind and rote obedience is no longer what many businesses require.[1] Businesses today require skills that extend well beyond those required of the relatively low skilled employees of the industrial age businesses. Today's businesses need people who can think for themselves, solve problems, discern what the problem might be, work as a member of a team, seek different points of view—explore options, and look for surprising connections—be open-minded when exploring possible solutions (Christensen, Horn, & Johnson, 2008; Conference Board of Canada n.d.; Gilbert, 2006; Wagner, 2008). Moreover, a predominate theme in much contemporary work on leadership in schools circles, not around pedagogical or curricular issues, but around managerial images and ideas, full of the very sorts of flow charts and assessment procedures shaped and formed in Taylor's work with industry, and all borne from the very distance from the day-to-day life of knowledge and its pursuit that defines the classroom that was typical of Taylor's image of shop managers[2] (Gilbert, 2005; Hargreaves & Fullan, 2009; Mintzberg, 2004; Senge, Scharmer, Jaworski, & Flowers, 2005; Wagner et al., 2006; Wagner, 2008).

This near-oxymoronic anti-intellectual aim of education was recognized over 80 years ago:

> The great H. L. Mencken wrote in *The American Mercury* for April 1924 [said] that the aim of public education is not:
>
>> To fill the young of the species with knowledge and awaken their intelligence. ....Nothing could be further from the truth. The aim...is simply to reduce as many individuals as possible to the same safe level, to breed and train a standardized citizenry, to put down dissent and originality. That is its aim in the United States...and that is its aim everywhere else. (cited in Gatto, 2006)

And now, and perhaps especially after eight years of the anti-intellectualism of the Bush administration, the narration-sickness of this tale persists in the work of teachers and students in many school settings. Attempts to resist such matters still feel its pull: when a science teacher in a local high school was recently told about a graduate course being offered to certain other teachers interested in addressing and thinking through the import of these matters, her response was "Gee, sounds nerdy!"[3]

One final fold of this tale for now. Our interested in F. W. Taylor's legacy emerged as what seemed to be a fold of the same cloth of our own work and its shadows. In *Back to the Basics of Teaching and Learning* (Jardine, Clifford & Friesen, 2008), we explore how the attractiveness of Taylor's industrial promise of efficiency dovetailed with the logic of fragmentation borne from a since-outdated version of the empirical sciences in the early 20th century. This dovetailed perfectly with the then-emerging Behavioral Sciences to produce an image of knowledge as built up one "basic" bit at a time ("the basics," in fact, became identified with those not-further-divisible "bits" out of which any knowledge was build, and "back to the basics" came to mean back to a version of knowledge-assembly right in line with Taylorian principles):

> The object is disassembled, the rules of its functioning are ascertained, and then it is reconstructed according to those rules; so, also, knowledge is analyzed, its rules are determined, and finally it is redeployed as method. The purpose of both remedies is to prevent unanticipated future breakdowns by means of breaking down even further the flawed entity and then synthesizing it artificially. Thus Gadamer [1989, p. 336] speaks of "the ideal of knowledge familiar from natural science, whereby we understand a process only when we can bring it about artificially." (Weinsheimer, 1987, p. 6)

The great affinity here is undeniable. 'Preventing unanticipated future breakdowns' is precisely the clarion of Taylor's efficiency movement, and deploying the rules of assembling in standardized ways was at its core.

This is the denouement in this old tale that once was full of early-20th-century enthusiasm. It is a tale that *purposefully designed to induce narration sickness*. If a student has a question, that means that they have a problem, and that problem then needs to be fixed so that there will no longer be any questions but simply obedience and compliance and its resultant productivities. If Taylor's dreams of efficiency thus come true, the classroom becomes set up such that *nothing happens* that is not anticipated and prescribed in advance. No surprise endings. As for a living discipline like mathematics and all the roiling work it requires, that life and work have no place in the efficient running of a mathematics classroom. More oddly put, the life of living disciplines is a problem for schools under the efficiency penumbra. Whatever cannot be efficiently surveilled and manageably delivered becomes erased from what is considered to be the 'field' of school(ed) mathematics. Mathematics becomes identified with its efficient rendering. The circle thus closes. Peter Cowley, author of The *Fraser Institute's Annual Report Card on Alberta Schools*, can thus speak to the issue of "teaching to the

[Provincially standardized] test [composed of questions wrought of such efficient rendering of mathematics]": "If teachers were following the provincial curriculum, by definition they would be teaching to the test. If they're not teaching to the test, then they're not doing their job." (cited in McGinnis, 2008, p. B5). No surprise endings, since now, of course, questioning such matters is acting unaccountably to provincial mandates and parental demands (the results of such provincial tests are published yearly in the local Calgary, Alberta newspaper under the auspices of 'parents' right to know,' without, of course, much discussion of what it is that is known in such knowing). Teachers and students (and University professors) who question such matters simply don't understand "the real world."

We end this meditation with a passage from Neal Stephenson's *Anathem* (2008, p. 414). It sets a fabled scene of fiction that is far more true to "the real world of schools" than those schools' claims of reality:

> Thousands of years ago, the work that people did had been broken down into jobs that were the same every day, in organizations where people were interchangeable parts. All of the story had been bled out of their lives. That was how it had to be; it was how you got a productive economy. If...employees came home at day's end with interesting stories to tell, it meant that something had gone wrong. The Powers That Be would not suffer others to be in stories of their own unless they were fake stories that had to be made up to motivate them.

Interlude

Many models of curriculum design seem to produce knowledge and skills that are disconnected rather than organized into coherent wholes. The National Research Council (1990, p. 4) notes that "To the Romans, a curriculum was a rutted course that guided the path of two-wheeled chariots. Vast numbers of learning objectives, each associated with pedagogical strategies, serve as mile posts along the trail mapped by texts from kindergarten to twelfth grade." (Bransford, Brown, & Cocking, 2000, 138). An alternative to a "rutted path" curriculum is one of "learning the landscape" (Greeno, 1991—tellingly for us, Greeno's paper is on numeracy). In this metaphor, learning is analogous to learning to live in an environment: learning your way around, learning what resources are available, and learning how to use those resources in conducting your activities productively and enjoyably (Greeno, 1991, p.175). Knowing where one is in a landscape requires a network of connections that link one's present location to the larger space. Traditional curricula often fail to help students "learn their way around" a discipline. (Bransford, Brown, & Cocking, 2000, p. 139)[4]

Thus the image of "learning the landscape" provides an alternative to the legacy of Taylor where curricular territory is, as the saying goes, "covered", replete with mile-posted breakdowns. This idea of a landscape in which one dwells, within which the relatedness and interdependence of those things which the fields sustains are essential to their well being, in which one's actions are those of obligation to and care for that well-being—all this echoes our own work for many years. We have explored mathematics as a sustaining field of relations that one must *inhabit* in order to understand, along with the perennial question of how to invite *this* class, or *that* particular student, into such "field knowledge."[5] We want to immediately reiterate that these seemingly subsequent pedagogical questions of invitation arise in a particular way. We are convinced that, as a living landscape or field, that field has within it a great range of diversity, multiplicity, modes and forms and figures. It has an elaborate ancestry of work and works, traces and tracks. As such, as a living field, mathematics is amenable to a wide range of explorers, a wide range of "learning styles" and interests, strengths and forgivable weaknesses, *because this is in the nature of a living field.* A living field *is* diverse and thereby has, in its nature as a living field, a breadth of embrace that shifts how we might discuss issues of diverse learning styles, English as a second language, special needs, multiple intelligences, learning delays, and the like. If we begin with image of mathematics that flows from Taylor's efficiency-fragmentation movement, with its requisite regimes of standardization, such diversity is a problem that must be fixed by multiplying the array of assembly lines to accommodate a "diversity of learners." We must simply state here that every teacher knows just how endless and burdensome this list can become if it is read against the background of fragmentation and efficient assembly. Under Taylor's shadow, everything accelerates and proliferates and scatters

Landscape, field, topography, a "larger space" of interdependent relations and kin, and of coming to know your way around such a field, and how the cultivation of memory is linked in with such matters—these commonplaces of hermeneutics (see Gadamer, 1989; Jardine, 2006a, 2006b)—sidestep this exhausting, often panic-stricken unraveling, acceleration, and proliferation. The time of the classroom changes from time always running out" to an oddly leisurely (L. *schola*, "leisure," ironically the root of the word "school") temporality that involves whiling (see Jardine, 2008; Jardine, Clifford, & Friesen, 2008; Jardine & Ross, 2009)—gathering, returning, exploring and slowly becoming experienced in the ways and relations of a place.

But these matters always sound so peculiar when spoken about in such general terms. Stephenson's meditation from *Anathem* provides a clue: an interesting story to tell is needed as a counterpoint to the now nausea inducing, standardized, amnesiatic, locked down character of the Taylorian tale of efficiency and the ways in which it tears apart the life of fields of knowledge (and the lives of those locked into this regime).

In the midst of ordinary classroom events, the experience of the opening of a living field of knowledge can be shown; it can be *experienced.* We'd like, now, to elaborate one such incident. As per the wont of hermeneutics, of course, when one talks about a field of relationships, any particular eventuality within that field—a student's question or concern, a particular set of experiences that students explore, a teacher's query, a particular diagram pulled from the pile of disposed-of work, and so on—is not simply to be read and recited as an isolated "anecdote," but is itself something that must be properly "fielded" in order to be properly, proportionally, "enfieldedly" understood. When treated as part of a living field of work, classroom examples are themselves not simply strung along Taylor's line of one-example-after-the-other to be statistically culled and deemed significant only to the extent that they frequently occur. Bluntly put, then, this idea of living fields of knowledge resists certain forms of research and eludes certain forms of scholarship that demand Taylor-like isolated cases as data. In hermeneutics, cases are not isolated but are experienced as "fecund" (see Jardine, 1992b). We must look into such examples for the fields they betray, and must, as effective teachers do with students' statements, questions, experiences, and explorations, do the work of fielding these—"reading" them out into the fields of living work in which they might belong, living fields which will house, exemplify, correct, adjust, make demands upon and cultivate such questions, experiences and explorations.

Fielding is thus a pedagogical practice. Differently put, understood as a living field, mathematics already involves generationality, intergenerationality, the young and the old, the new and the established. As a living field, the continuance of that life and the conditions of that continuance are not an afterthought but at the heart of the matter.

Treated as a living field, mathematics *is* pedagogical.

"AT DAY'S END WITH INTERESTING STORIES TO TELL"

> *There is no art or technique of happening onto things. There is no method of stumbling.*
> —Weinsheimer, 1987, p. 7

There is a wonderful etymological twist that is at work here. To become experienced means "to learn your way around," that is, to have ex-*peri*-ence (Gk. *ek*- means "out of," *peri*- means "around, as in the term "perimeter"—the "measure" [*metron*] of "around" [*peri*]). (Jardine, Friesen, & Clifford, 2006, p. 207)

We are interested in the experience of mathematics as a field, how it becomes a place that can allow learning one's way around. There is no technique precisely to such an experience other than experience itself. Pedagogically, it is a matter of setting students out into this field and setting up experiences that make field experience possible.

Groups of four Grade Seven students were given pencils, large sheets of paper, a straight edge and a compass, and were given two tasks:

- Construct a perpendicular line from a point above that line and
- Bisect an angle.

These tasks, of course, sound stunningly and numbingly familiar to any one who has been in school. We could each rattle off the Taylor-like rules in each case. The *Internet* both helps and hinders. Here is a clear, mindlessly cut-and-pasted set of such rules found in under a minute's exploration:

> To drop a perpendicular from a point to a line:
>
> > Given a line and a point A not on the line open the compass a distance larger than the perpendicular distance from the point to the line, and scribe an arc intersecting the line at two points.
> >
> > Call the two points B and C. Scribe arcs from B and C of the same radius on the other side of the line from A.
> >
> > Call the point where the two arcs meet D. Draw AD.
> >
> > AD is the line through A perpendicular to BC. (Sonoma, 2009)

Most of the students and teachers and student teachers we encounter have all been schooled in a version of mathematics where one memorizes such rules or procedures or operations and how they work, then one is give problems and practices applying one to the other. Then one is give "home-work" where you repeatedly do such problems. Then you are tested to see if you have 'mastered' these matters, and then this process is simply repeated for the next chapter, unit, grade and so on. Instead, these students were given the instruments available in ancient Greece and were told to find out how to accomplish the tasks assigned. So it was difficult, nearly ten years ago now, to watch groups of Grade Seven students struggle through these tasks. It was agonizing to look over shoulders and know the memorized rules and remain silent.

We cannot reproduce all of the ins and outs, all of the arguments and diagrams and scrawled attempts. Suffice it to say, for now, that, after quite a long series of attempts, one of the groups of boys finished an example of dropping a perpendicular line. Figure 10.1 is a version of what they did.

Strangely familiar to any of us who've been schooled—that cross-hatch of 2 arcs of 2 circles below the line providing the point needed.

An aside: if all we wanted this group of students to accomplish was to remember a usable version of the steps to go through, such an experiential

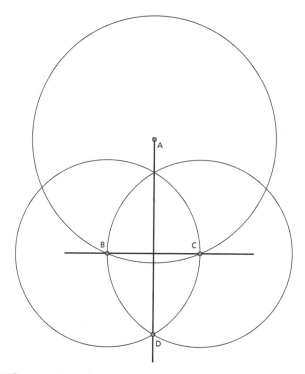

Figure 10.1 Construction of perpendicular to a line segment.

and exploratory task is clearly an inefficient waste of time. Some teachers we've encountered will get this far along: giving students the task of treading this weird field, experimenting and so on, but often we find that, once the "activity" is done, students are asked to generate a list of rules like the one above, are then given textbook questions to practice and look up answers in the back of the book, and they are then tested on their mastery of this procedure. As we've found over decades of suffering with students and teachers, students who understand what is happening here will, the next time an activity is set up by the teacher, simply ask: "Why don't you just tell me how to do it?" Experiences and explorations such as this come to be seen as pedagogically viable *vehicles* for 'getting across' what is, in fact, simply a straightforward set of steps that we could simply tell to students and have them practice. Frankly, I (D.J.) felt a bit like this walking around the classroom. I realized something about how concrete experimentation and argument and the like were good ideas, but what I didn't understand was how dropping a perpendicular or bisecting an angle was understandable as anything *more than* simply being able to follow the rules—being able to "do it", since that is all that examinations seemed to test. This is an important place of being stuck that is rampant and endemic in schooling and its results: mathematics is simply the rules, and not a living, experiencable, pleasurable, field. That feeling of being closed in when in school(ed) mathematics.

These rules apply to that.

Those to this.

Don't forget!

And, of course, there is nothing especially memorable about those rule lists. They don't require anything except memorization. They are not memorable. There is an interesting, memorable story to tell about these matters only if something now goes wrong.

When repeated, these rules are monotonously always exactly the same, standardized like Taylor's assemblies. The more examples I practice, the less interesting this whole enterprise becomes, and the command of the rules becomes simply automatized. I become ejected from that field, dejected. My initiative, as was Taylor's wont, will only mess things up.

The next task, bisecting an angle, also has rules that our hands and memories have housed and that don't bear repeating right now. One group of students put aside their multiple, scrawly drawings of a perpendicular-drop (which, in the event, was full of false starts, erased lines, scrawled notes and the like) and began to take on the next task. I've (D.J.) had the great pleasure over many, many years of finding out that something more is in store than simply the next set of rules to be learned. If all Sharon wanted was for them to memorize these two sets of rules and be able to master applying them to examples like on the Provincial Examinations, she would have just told them. These students were being drawn into a world where:

> [S]omething is going on (*im Spiele ist*), something is happening (*sich abspielt*)" (Gadamer, 1989, p. 104) [German: *Spiel,* "play"—something is "at play"]. Students were being drawn into an abundant and venturous play-space (*Spielraum*) that was full of ancestors, ghosts, inheritances, bloodlines, comforts, faces, dates and names and kindred spirits, ongoing contestations and conversations that are still happening. It is precisely this sense of entering a world larger than ourselves that portends the freedom and ease experienced in playing. (Jardine, Clifford & Friesen, 2006, p. 59)

There is an experience being hinted at here that is not reducible to simply knowing the rules and that is not actually even hinted at when one only knows the rules. These students aren't being put through this work simply in order to discover the rules you could have simply told them. They are being put through this as a way to open up and start to make available and experienceable the field(s) within which angles, points, right angles, circles and the like live. We know, at this juncture, how Taylor would respond to a venture that "has some play in it."

What is this "field" that is being portended here?

D. J. happened to be at one end of a table of four boys doing these tasks. They had cast eight or ten pieces of newsprint from the perpendicular exploration off the table to the side and were looking back at the instructions: bisect an angle.

They started off, and it is hard to remember exactly how long things went on as with the earlier part of the task before one of the boys said, "Stop, stop. We've done this already."

There was, for me (D.J.) at least, the weirdest initial moment of abstract pedagogical pleasure: good, something is happening, conversations are going to rage and quell, possibilities are going to be opened up. We both admire these and contend that they are at the heart of generous and sustainable pedagogy. But I really had no idea what this student was talking about. We've done this already? No. *That* was dropping a perpendicular. *This* is bisecting an angle.

This student rummaged through the pages that had been discarded. He picked up one of the perpendicular-drop diagrams and turned it sideways (Figure 10.2).

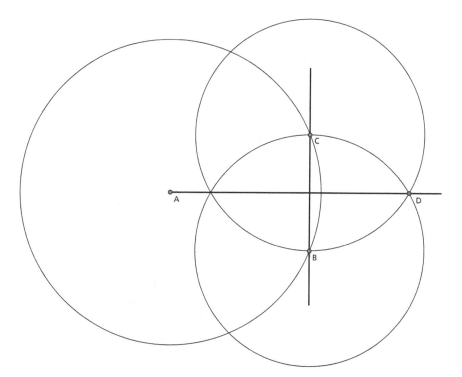

Figure 10.2 Sideways orientation of Figure 10.1.

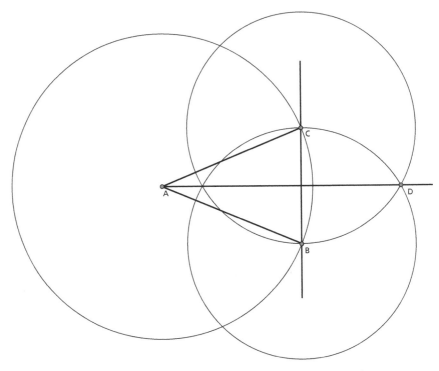

Figure 10.3 Relationship between bisecting an angle and constructing a perpendicular line.

He and his group then drew line segments AC and AB (see Figure 10.3) as his fellow students were a mixture of argumentative, bewildered, confused and amazed, "We've already done this."

The line AD which was heretofore understood as a hard won perpendicular, suddenly bisects the angle CAB.

Now there is a phenomenological experience here that needs noting, despite its ephemeral nature: that right in the midst of a seemingly "completed diagram" of a dropped perpendicular, a set of as-yet-unexplicated or explored relations, consequences, commonalities, resemblances was portended. Slowly emerging was that the first piece of work was, so to speak, not simply somehow self-containedly "itself" (one piece on Taylor's mathematical assembly line) but was what properly itself only within an emerging set of fields of relations. What was initially experience as a particular piece of work slowly starts to become experiencable as an "opening" into a field of heretofore hidden relations. Dropping a perpendicular line seems to exist in a field in which it has "relatives" and "kin." Dropping a perpendicular line and bisecting an angle live in the same field of relations. They are "nearby."

Now there is another aspect to this, shall we say, "fielding phenomenon," one of "draw." As Hans-Georg Gadamer suggests regarding the work of art, so, too, here: "[it] compels over and over, and the better one knows it, the *more* compelling it is. There comes a moment in which something is *there*, something one should not forget and cannot forget. This is not a matter of mastering an area of study" (Gadamer, 2007, p. 115). Rather, we come to experience "a world into which we ourselves are drawn" (Gadamer, 1994, p. 192). As that first diagram shifts sideways, and attention is continued upon that cross hatch of two intersecting circles that provides the point of perpendicularity and the point of angular bisection, as attention continues to be cultivated, the draw of this world of relations increases rather than decreases.

This is something central to the pedagogical point here. Coming to know one's way around such matters results in an *increase* of the experience of the compelling character of the "place"/field one has become experienced in. Moreover, as such experience of the "there" of this expanding field increases, we become more and more susceptible to new possibilities of experience:

> "Being experienced" does not consist in the fact that someone already knows everything and knows better than anyone else. Rather, the experienced person proves to be, on the contrary, someone who ... because of the many experiences he has had and the knowledge he has drawn from them, is particularly well-equipped to have new experiences and to learn from them. Experience has its proper fulfillment not in definitive knowledge but in the openness to experience that is made possible by experience itself. (Gadamer, 1989, p. 355)

Students in that Grade Seven class, on gathering around these sideways diagrams, start to ask "What is going on?" Out from under Taylor's easily memorizable rules, something memorable is forming, something memorably *mathematical.* In the vertigo draw of that turned diagram and the rushing intake of breath and an arriving insight, we recognize mathematics as a living field that we find ourselves inhabiting.

OPENNESS TO EXPERIENCE

Field knowledge surrounds the particulars of that field with penumbras of sense and significance and relatedness. Mastery over that field no longer means surveillance and management, but knowing one's way around that field and being able to venture into it and have new experiences of its ways. All the rules are still there, but this "there-ness" has changed in ways difficult to describe. The curriculum is still "covered" but no longer simply cover over. There is memorability here, not just memorization.

There is another thread to this tale that started around what we've called "A Secret Meeting with Pythagoras' Ghost" (Jardine, Clifford, & Friesen, 2006, p. 1–3). We recounted there an old story of a 12 year old boy who was part of those dropped perpendiculars and part of those angle bisection investigation. Out on a playground, near noon, winter, facing south, with his toes near the top of a shadow of a tree straight south. Talking about how long the shadows are, and how much of the playground was sunlit in summer when the sun was higher. His offhand comment, "but Pythagoras says that something is *the same*."

Jump ahead 10 years, where the swirls of mathematical memorability are still vivid. Playing around with knowledge that was never found in our own schooling, that every instance of two numbers multiplied together in fact describes a geometrical *area*. So it seems that the Pythagorean Theorem could be pictured as in Figure 10.4 (note that we have shifted the labeling of this diagram from the figures above), where the area of squares 1 plus 2 is equal to the area of square 3.

Simple.

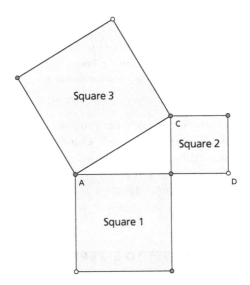

Figure 10.4 Pythagorean 3–4–5 triangle.

Then one of those unnecessary, opulent, frivolous thoughts, that the sides of such a triangle are line segments, just as are the radii of circles. And then, yes, in the formula for the area of a circle, a line segment of a certain length—the circle's radius— is *squared*: πr^2. So if the Pythagorean Theorem states that $a^2 + b^2 = c^2$, it seems that each of these line segments (a, b and c) could easily be thought of as *the radii of three circles*. Therefore, this might work: $\pi a^2 + \pi b^2 = \pi c^2$, such that the area of a circle of radius a, plus the area of a circle of radius b would equal the area of a circle of radius c.

But what would happen if we drew this (Figure 10.5)?

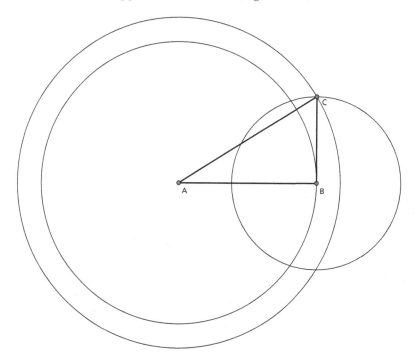

Figure 10.5 Pythagorean 3–4–5 triangle where $\pi a^2 + \pi b^2 = \pi c^2$.

As happens with field knowledge, this is starting to look familiar. We're getting the feeling that we know this field. Haven't we "done this already?"

By doubling the Pythagorean Triangle within the large circle with radius AC, something old and familiar started to appear all over again (see Figure 10.6).

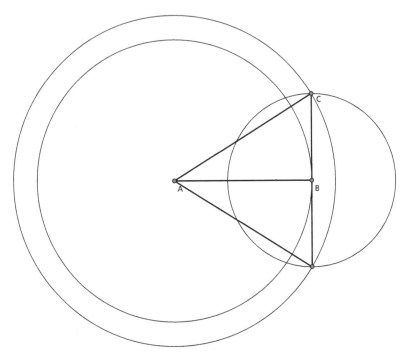

Figure 10.6 Relationship between $\pi a^2 + \pi b^2 = \pi c^2$, bisecting an angle and constructing a perpendicular line.

Now where are we? Back into the same field we thought we had quit. But why are we back here in this field again? What is it we missed when whiling over that first student diagram that was breathtakingly turned sideways? These things are *related?* So again, just at the juncture where the perpendicular and bisecting constructions were on the verge of turning into mere objects to be manipulated, both of them "field" again.

More elaboration is shown in Figure 10.7.

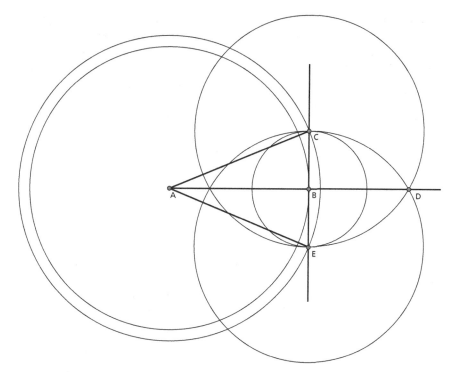

Figure 10.7 Relationship between Figure 10.6 and Figure 10.3.

But isn't this Figure 10.3?

ENDBIT

[Hinted at here is] the sense of "nativeness" ["knowing your way around"], of belonging to the place. Some people are beginning to try to understand where they are, and what it would means to live carefully and wisely, delicately in a place, in such a way that you can live there adequately and comfortably. Also, your children and grandchildren and generations a thousand years in the future will still be able to live there. That's thinking as though you were a native. Thinking in terms of the whole fabric of living and life. (Snyder, 1980, p. 86)

We'll leave readers to figure these figures further and to ask themselves questions of the pedagogy of such matters. In the face of this suddenly emerging Pythagorean ghost, if any of us gets bewildered by the reappearance of an old image from discarded drawings on the floor, if we feel the vertiginous sense of a crowd of familiar images suddenly giving way to what Heidegger (1962, p. 171) called a movement of *clearing*—an opening space into which the familiar is then set loose (think of how, in the face of the Pythagorean diagrams above, dropping a perpendicular becomes set out anew into an open landscape, how, in fact, just as with an ecological space, one species invokes another's presence)—this is not a *problem*, this is not *inefficiency*, even though this is precisely the sort of field experience that Taylor's work, in its application to schooling, purged from the mathematics classroom. It is *fielding* knowledge. It is mathematics experienced as a "landscape," as a living field, a "living discipline."

It is this sort of experience of being drawn into a living field of work and having that field ask something of us that is at the heart, we suggest, of a viable pedagogy. "Something asked of us" resists the exhaustion of a constructing subjectivity cut loose in a splay of fragments.

Such living fields don't need management, nor do they need the ecologically disastrous belief that "we make all the patterns" (Berry, 1987, p. xv). They need only a certain continuity of attention and devotion and the slow gathering of experience ("knowing your way around") that is the gift of such devotion.

It is on behalf of precisely such a draw into a living field that our work has been directed and that we give thanks for the example that Professor Ernest has provided over these many years of work.

NOTES

1. In the most recent (January 27, 2009) Canadian federal budget, it was decided that the agency that has heretofore provided research funding for the discipline of education (the Social Sciences and Humanities Research Council of Canada [SSHRCC]) become henceforth "focused on business-related degrees" (see The Undersigned, 2009; see also Church, 2009 and countless other responses on line).

2. Mintzberg (2004, pp. 238–241) extracts a section of Clifford & Friesen, 1993 article A curious plan: Managing on the twelfth to signal the change needed in management education.

3. In a 2003 article from *Canadian Living Magazine* entitled "Heartbreak Joy," (Harrison 2003), we find echoes. This article is a popular consideration of the reasons for the increase in instances of autism and asks:

   Does better diagnosis account for the increase? Not entirely. There is some evidence that autism may be the result of an unhappy constellation of genes. The emergence of what some people call the "geek connection" may back this up: an expanded computer industry is creating more opportunities for two people with special aptitudes for math and sciences to get together, explains Jeanette Holden, the program director of autism spectrum disorders at the Canadian-American Research Consortium in Kingston, Ont. Although they probably don't have autism themselves, these computer programmers might have one or more of the rarer "ASD [Autism Spectrum Disorder]-like" traits, such as extraordinary memory capabilities. The offspring of such parents, says Holden, may inherit a "double dose" of these ASD-like traits, which may predispose them to ASD. (p. 72)

We wait, of course, for the eugenics to fully engage, and for engaging real and lively mathematics conversations in schools to be blamed for triggering a latent genetic defect.

4. It is important to note that this text, as well as Sawyers 2006 make no reference to the reconceptualist movement and its work on *currere*, which dates back to 1975 and which links to the philosophical movements of phenomenology, critical theory and hermeneutics. This accounts, we suggest, for one of the reasons why our own work has been often considered oddly radical and marginal, and "tinged with hyperbole." Although it is a positive sign to see this idea of "landscape" becoming legitimized, its ahistoricality is rather disturbing (Aoki, Pinar, & Irwin, 2004; Clifford & Friesen 1993; Doll 1993; Jardine 1990, 1992b, 1994; Pinar 2003).

5. It should be noted that many attempts at "reform" in education have recently taken for granted the fragmentation presumed by Taylor and simply tossed in a "constructing subjectivity"—what Jean Piaget (1971, xii) called the child's "imposing cosmos on the chaos of experience." Regarding the dangers of such "constructivism," see Jardine 2005.

REFERENCES

Aoki, T. (author), Pinar, W., & Irwin, R. (Eds). (2004). *Curriculum in a new key: The collected works of Ted T. Aoki.* Mahwah, NJ: Lawrence Erlbaum.

Ayres, L. (1909). *Laggards in our schools.* NY. Retrieved March 27, 2009 from http://www.archive.org/details/laggardsinoursch00ayrerich

Ayres, L. (1915). *A measuring scale for ability in spelling.* Retrieved July, 20, 2009 from http://www.donpotter.net/PDF/AyresSpellingScale.pdf.

Berry, W. (1986). *The unsettling of America: Essays in culture and agriculture.* San Francisco: Sierra Club Books.

Berry, W. (1987). *Home economics.* San Francisco: North Point Press.

Boyle, D. (2006). The man who made us all work like this.... *BBC History Magazine,* June 2003. Retrieved March 27, 2009 from http://www.david-boyle.co.uk/history/frederickwinslowtaylor.html

Bransford, J., Brown, A., & Cocking, R. (Eds) (2000). *How people learn: Brain, mind, experience and school.* Washington, DC: National Academies Press.

Callahan, R. (1964). *America, education and the cult of efficiency.* University of Chicago Press.

Christensen, C., Horn, M., & Johnson, C. (2008). *Disrupting class: How disruptive innovation will change the way the world learns.* NY: McGraw-Hill.

Church, E. (2009). *Scholarships for business studies draw outrage.* February 20, 2009. Retrieved March 27, 2009 from http://www.globecampus.ca/in-the-news/article/scholarships-for-business-studies-draw-outrage/

Clifford, P., & Friesen, S. (1993). A curious plan: Managing on the twelfth. *Harvard Educational Review 63*(3), 339–358.

Clifford, P., & Marinucci, S. (2008). Testing the waters: Three elements of classroom inquiry. *Harvard Educational Review 78*(4), 675–688.

Conference Board of Canada (n.d.). *Innovation skills profile.* Retrieved March 27, 2009 from http://www.conferenceboard.ca/topics/education/learning-tools/isp.aspx

Cubberley, E. P. (1922). *A brief history of education: A history of the practice and progress and organization of education.* Boston: Houghton Mifflin.

Doll, W. (1993). *A post-modern perspective on curriculum.* NY: Teachers College Press.

DuFour, R., & Eaker, R. (1998). *Professional learning communities at work: Best practices for enhancing student achievement.* Bloomington, IN: Solution Tree.

Friesen, S., Clifford, P., & Jardine, D. (1998). Meditations on community, memory and the intergenerational character of mathematical truth. *Journal of Curriculum Theorizing 14*(3), 6–11.

Friesen, S., Clifford, P., & Jardine, D. (2003). Jenny's shapes. *Philosophy of Mathematics Education Journal* An on-line journal. Retrieved March 27, 2009 from http://www.ex.ac.uk/~PErnest/pome17/jenny.htm

Gadamer, H.-G. (1989) *Truth and method.* J. Weinsheimer, trans. NY: Crossroads.

Gadamer, H.-G. (1994). *Heidegger's ways.* J. W. Stanley, trans. Boston: MIT Press.

Gadamer, H.-G. (2007). From word to concept: the task of hermeneutics as philosophy. In R. Palmer, (Ed.), *The Gadamer reader: A Bouquet of the later writings,* (108–122). Evanston, IL: Northwestern University Press.

Gatto, J. (2006). *The national press attack on academic schooling.* Retrieved March 27, 2009 from http://www.rit.edu/~cma8660/mirror/www.johntaylorgatto.com/chapters/9d.htm

Gilbert, J. (2005). *Catching the knowledge wave? The knowledge society and the future of education.* Wellington, NZ: NZCER Press.

Greeno, J. (1991). Number sense as situated knowing in a conceptual domain. *Journal for Research in Mathematics Education 22*(3), 170–218.

Hargreaves, A., & Fullan, M. (Eds.) (2009). *Change wars.* Bloomington, IN: Solution Tree.

Harrison, P. (2003). Heartbreak joy. *Canadian Living Magazine (March),* 69–78.

Heidegger, M. (1962). *Being and time.* NY: Harper & Row.

Hillman, J. (1989). *Healing fiction.* Barrytown, PA: Station Hill Press.

Jardine, D. (1990). On the humility of mathematical language. *Educational Theory, 40*(2), 181–192.

Jardine, D. (1992a). *Speaking with a boneless tongue.* Bragg Creek, Alberta, Canada: Makyo Press.

Jardine, D. (1992b). The fecundity of the individual case: Considerations of the pedagogic heart of interpretive work. *British Journal of Philosophy of Education. 26*(1), 51–61.

Jardine, D. (1994). The ecologies of mathematics and the rhythms of the Earth. In P. Ernest, (Ed.), *Mathematics, philosophy and education: An international perspective, studies in mathematics education, Vol. 3*, (pp. 109–123). London: Falmer Press.

Jardine, D. (1995). The stubborn particulars of grace. In B. Horwooded (Ed.), *Experience and the curriculum: Principles and programs*, (pp. 261–275). Dubuque, Iowa: Kendall/Hunt.

Jardine, D. (1997). Under the tough old stars: Pedagogical hyperactivity and the mood of environmental education. *Clearing: Environmental Education in the Pacific Northwest 97*(April/May), 20–23.

Jardine, D. (2000). *Under the tough old stars: Ecopedagogical essays*. Brandon, VT: Psychology Press/Holistic Education Press.

Jardine, D. (2004). *Speaking with a boneless tongue. 2nd Edition*. Retrieved July 20, 2009 from http://www.ex.ac.uk/~PErnest/pome16/boneless_tongue.htm.

Jardine, D. (2005). *Piaget and education: A primer*. NY: Peter Lang.

Jardine, D. (2006a). Youth need images for their imaginations and for the formation of their memories. *Journal of Curriculum Theorizing 22*(3), 3–12.

Jardine, D. (2006b) On hermeneutics: What happens to us over and above our wanting and doing. In K. Tobin & J. Kincheloe (Eds.), *Doing educational research: A handbook*, (pp. 269–288). Amsterdam: Sense.

Jardine, D. (Feb, 2008). On the while of things. *Journal of the American Association for the Advancement of Curriculum Studies*. Retrieved March 27, 2009 from http://www.uwstout.edu/soe/jaaacs/vol4/Jardine.htm

Jardine, D., Clifford, P., & Friesen, S. (2008). *Back to the basics of teaching learning: Thinking the world together*, 2nd ed. NY: Routledge.

Jardine, D., & Friesen, S. (1997). A play on the wickedness of undone sums, including a brief mytho-phenomenology of "x" and some speculations on the effects of its peculiar absence in elementary mathematics education. *Philosophy of Mathematics Education Journal 10*. Retrieved July 20, 2009 from http://people.exeter.ac.uk/PErnest/pome10/art6.htm.

Jardine, D., Friesen, S., & Clifford, P. (2003). Behind each jewel are three thousand sweating horses: Meditations on the ontology of mathematics and mathematics education. In E. Hasebe-Ludt & W. Hurren (Eds.), *Curriculum intertext: Place/language/pedagogy*, (pp. 39–50). New York: Peter Lang.

Jardine, D., Friesen, S., & Clifford, P. (2006). *Curriculum in abundance*. Mahwah, NJ: Erlbaum.

Jardine, D., & Ross, S. (2009). Won by certain labour: A conversation on the while of things. *Journal of the American Association for the Advancement of Curriculum Studies*. Retrieved July 23, 2009 from http://www.uwstout.edu/soe/jaaacs/Vol5/Ross_Jardine.htm.

Kanigel, R. (2005). *The one best way: Fredrick Winslow Taylor and the enigma of efficiency*. Cambridge MT: The MIT Press.

McGinnis, S. (2008). Province urged to scrap high-stakes tests. *Calgary Herald* (Sept., 8), B5.

National Research Council. (1990). *Reshaping school mathematics*. Mathematical Sciences Education Board. Washington, DC: National Academy Press.

Piaget, J. (1971). *The construction of reality in the child*. New York: Ballantine Books.

Pinar, W. (2003). *International handbook of curriculum research (Studies in curriculum theory)*. New York: Routledge.

Rodgers, G. (1983). *To urge the repetition of the Ayres' spelling tests of 1914–15 to confirm the existence of massive present-day reading disability and to establish its cause and cure*. Retrieved July 20, 2009 from http://www.donpotter.net/PDF/Ayers-ToUrge.pdf.

Rodgers, G. (1984). Historical introduction to Leonard P. Ayres' *A measuring scale for ability in spelling (1915)*. Prepared by Donald L. Potter, July 21, 2004, from materials written by Geraldine Rodgers on December 30, 1984. Retrieved March 27, 2009 from donpotter.net.

Sawyer, R. K. (Ed) (2006). *The Cambridge handbook of the learning sciences*. New York: Cambridge University Press

Senge, P., Scharmer, C. O., Jaworski, J., & Flowers, B. (2005). *Presence: An exploration of profound change in people, organizations and society*. New York: Currency.

Smith, D. G. (1999). Brighter than a thousand suns: Pedagogy in the nuclear shadow. In D. G. Smith (1999). *Pedagon: Interdisciplinary essays in the human sciences, pedagogy and culture*, (pp. 127–136). New York: Peter Lang.

Snyder, G. (1980). *The real work*. New York: New Directions Books.

Sonoma (2009). *Construction #4: To drop a perpendicular from a point to a line*. Retrieved March 27, 2009 from http://www.sonoma.edu/users/w/wilsonst/courses/math_150/c-s/Drop-Perp.html

Stephenson, N. (2008). *Anathem*. NY: Harper Collins.

Taylor, F. W. (1903). *Shop management [Excerpts]*. Retrieved March 27, 2009 from http://www.marxists.org/reference/subject/economics/taylor/shop-management/abstract.htm

Taylor, F. W. (1911). *Scientific management, comprising shop management, the principles of scientific management and testimony before the special house committee*. New York: Harper & Row.

(The) Undersigned. (2009). *The business of scholarships: An open letter to parliamentarians*. February 4, 2009. Retrieved March 27, 2009 from http://www.rabble.ca/news/business-scholarships-open-letter-parliamentarians.

Wagner, T. (2008). *The global achievement gap: Why even our best schools don't teach the new survival skills our children need—and what we can do about it*. New York: Basic Books.

Wagner, T., Kegan, R., Lahey, L., Lemonds, R., Garnier, J., Helsing, D., Howell, A., & Rasmussen, H. (2006). *Change leadership: A practical guide to transforming our schools*. San Francisco: Jossey-Bass.

Weinsheimer, J. (1987). *Gadamer's hermeneutics*. New Haven, CT: Yale University Press.

Wiebe, R., & Johnston, Y. (1998). *Stolen life: The journey of a Cree woman*. Toronto: Alfred Knopf Canada.

Wittgenstein, L. (1968). *Philosophical investigations*. Cambridge UK: Blackwell's.

Wrege, C. D., & Greenwood, R. (1991). *Frederick W. Taylor: The father of scientific management : Myth and reality*. NY: Irwin Professional. The text of *Chapter 9* retrieved March 27, 2009 from johntaylorgatto.com/chapters/9d.hFtm

CHAPTER 11

哲学の道

Paul Dowling
Institute of Education, University of London

"Good morning, Paul (yes, you, not me; I rarely greet myself in this way if, indeed, at all). I find myself at 白河今出川の交差点, the eastern end of 哲学の道 in 京都. It's a glorious spring day. 桜 is in full bloom—a pink and white aria; a baseline of black bark; a sky blue chorus—embarrassing the sullen 鯉 in the 琵琶湖疏水 below (their own reflections are on more grubby things). I'm going for a walk; perhaps you'll join me. We're not alone, of course. Our companions—in their hundreds, I guess—are, like me, pointing their cameras at the flowers, at the fish, at each other. The cameras range from high-end professional jobs, like mine (this is a boast), thru midrange SLRs and micro digicams to mobile phones with megapixel ratings that quite possibly outrank my EOS 1D Mark III, though not, I think, my 5D Mark II (another one). The snaps flatten our walk in more ways than one, most tragically, perhaps, the flowers no longer sing in the light breeze; they are caught agape. Even the high-definition video facility of the 5D Mark II flattens, though in a different way: where is the warmth of the sun on my back (we snappers generally face away from the sun)? And, anyway, on another walk, on another day, the music will be different.

"I should have explained, 哲学の道 (*tetsugaku no michi*)—starting at the junction between *Shirakawa dori* and *Imadegawa dori* in *Kyoto*—is generally

Relatively and Philosophically Eᵃrnest, pages 173–190

rendered in English as 'the philosopher's walk.' It follows a narrow canal that is lined with 桜 (*sakura*, cherry trees) and home to some oversized and rather lugubrious koi. The path is named for the daily meditational walks of the Kyoto University philosopher, 西田 幾多郎 (*Nishida Kitarō*, 1870–1945). I find Nishida's philosophy, insofar as I understand it from my brief research, rather appealing and not entirely unconnected with our walk in a conceptual way, but further exploration is perhaps best left for another walk, another day. Today, I'd like to start our walk with a quote from André Gide: 'Everything has been said before, but since nobody listens we have to keep going back and beginning all over again.' Well, I think I know how he might have felt, but, to be blunt, everything hasn't already been said before and even if it has, saying the same thing on different occasions, in different contexts is saying something else. Not only that, but listening isn't an issue either, unless it is constituted as a potentially perfect reception and that's pie in the sky. So, everything may or may not have been said already, but everything nevertheless remains to be said.

"The key to all of this, in my formulation, is recontextualisation. I recall that you have made very productive use of this term in a paper on semiotics and mathematics published not so long ago (Ernest, 2006). I noticed that, in this paper, you made reference to an old article of mine (Dowling, 1989, 1991a).[1] The title of my paper is relevant to our walk, as will become clear: *The Contextualising of Mathematics: Towards a Conceptual Map.* In the paper I had attempted to construct a map of the contexts in which mathematical activity takes place. The map was very primitive. It seemed to propose that, for example, we could identify in some sociologically significant way, fields of production and recontextualisation, reproduction and acquisition of mathematical knowledge that roughly corresponded to the university, textbook and curriculum production, teacher and learner activity. I had rather distorted some of the ideas of Basil Bernstein—my mentor at the time. In your paper you constructed a resonant schema, but constituted as articulated semiotic systems as opposed to sociological fields, your schema identified transformations between mathematical theory, school mathematics topic, taught school mathematics topic, and mathematical topic as mastered by the learner. Now, shearing away the sociological (the category 'field' implicates agents as well as their practices) enabled you to deal with situations that presented theoretical conundrums for me. An example would be that 'most school mathematics topics are no longer part of academic (university) mathematics and thus figure in no contemporary academic textbooks' (Ernest, 2006; p. 73). Sociologically different contexts can certainly be described as recontextualising their respective practices, but historical transformations also effect recontextualisations; fields must be diachronically as well as synchronically contextualized. This being the case, I could not constitute the fields in my schema with any sociological integrity; not, that is, without constructing a map almost as complex as the 'reality' of the twenty to fifty thousand years (I think you implied) of the 'real' history of the mathematics curriculum.

"Semiotics is, of course, about meaning and so is able to background social relations, allowing a focus on aspects or dimensions of semiotic systems. This is exactly what you did in your paper, presenting analyses of historical, mathematical and developmental perspectives on number, counting and computation, without needing to claim exhaustion of the terrain, which, in a sense, the field approach must do, albeit at a necessarily high level of abstraction. The resulting three-dimensional view of mathematics presents an interesting and, I think, pedagogically useful resource. Your analysis also emphasises diachronic as well as synchronic texture. Does it, though, escape the flattening effect of the cameras on the philosopher's walk? Unfortunately, you are not in a position to answer right now and I will also defer my own answer to my own question. However, I must admit that I have frequently suppressed the diachronic in my analysis. Let me show you what I mean with some examples; I hope you won't mind if they tend to draw on illustrations that are not directly mathematical; the results are, in my view, all directly relevant to mathematics education.

"It seems that books made by monks in mediaeval scriptoria were a long way from calligraphic perfection. In one 'school for scribes,' Cohen-Mushlin (2008), for example, has described the way in which the master would scribe a few lines to act as a model for his (sic) pupil, who would then take over the production of the book. Almost inevitably, however, the pupil's work was not up to scratch and the master would produce another exemplar. This might be repeated many times. As the pupil progressed, they would move to more challenging tasks, such as rubrication, again following the master's exemplars. Eventually, the pupil might himself (sic) become a master, but, in the meantime, the book that he had been working on would be finished, complete with all of his inadequate work and, quite probably, the inadequate work of other pupils, together with the (presumably) adequate work of various masters. Parchment was too costly and the time involved in scribing too extended to allow pupils to practice to perfection before putting their marks on what would be the final (and first) version of a book. The result would be something of a hotchpotch in terms of calligraphy.

"This is very different from the situation that obtains in respect of the apprenticeship of Japanese 民芸 potters (see Singleton, 1989). The apprentice, on initial entry to the pottery, is not allowed any involvement in the production process for quite an extended period of time, being limited to marginal activities, such as cleaning and making tea and, of course, watching. Eventually, the novice is allowed onto the wheel and told to make 一万の酒の杯 (ten thousand sake cups). Naturally, the apprentice's initial attempts are unworthy and so consigned to the bin to be recycled. Much later, some of his attempts will receive approval and be fired and sold in the shop—as seconds.

"Both of these modes of pedagogy take place in the context of production, which is to say, the scriptorium and the potter's workshop are the sites of the productive practices to which the novices are being apprenticed and their teachers are recognised practitioners of their respective crafts. However, there seems to be a fundamental difference. In the pottery, we have what looks like a traditional apprenticeship that, indeed, resembles Lave and Wenger's (1991) 'legitimate peripheral participation' model. The apprentice is not allowed to be involved in actual production until their performance reaches a satisfactory standard; nothing will go out of the pottery unless it meets this standard. In the scriptorium, on the other hand, it is not possible to wait until the pupil reaches mastery. Nevertheless, pedagogy continues during the productive process. Emphasis, here, seems to be more on the production of a community of competent practitioners; like the ideal of a perfect product, this is also unachievable to the extent that there will be a continuous influx of new novices. The distinction, then, is between the prevalence of strategies that stress competence—the scriptorium—and the prevalence of those emphasising performance—the pottery.

"Not all pedagogy takes place in the context of production. In the high school, for example, the nominal school subject is generally mediated by a curriculum and by a teacher, whose principal expertise is in teaching rather than in the discourse that they are relaying. The two dimensions of pedagogy that I have now introduced—what I shall call, 'transmitter focus' and 'mediation'—generate this strategic space (Figure 11.1).

"The construction of this space has involved looking—albeit in secondary and non-systematic ways—into an empirical space that reaches from mediaeval Germany to present day Japan, the various settings analysed by Lave and Wenger and, through my empirical observation on schooling, the UK. This is a very substantial chronotope that would defy an attempt at a totalising sociology. My schema, however, has flattened and constricted the chronotope in the construction of a perspective. The potential sociologies of the various settings that constitute the chronotope have been ignored or, rather, reduced to fundamental strategies in a move perhaps similar

	Transmitter Focus	
Mediation	Competence	Performance
Unmediated	*delegating*	*Apprenticing*
Mediated	*teaching*	*Instructing*

Figure 11.1 Transmission strategies from Dowling (2008).

to your semiotic move. However, the schema is now available as a logically complete matrix of strategic possibilities from the perspective generated by taking the cross-product of transmitter focus and mediation. The central categories in this matrix—delegating, apprenticing, teaching, and instructing—do not totalise any pedagogic practice. We may reasonably speculate that any empirical transmitter will recruit more than one and possibly all of these strategies. For my present purposes, some explication of the terminology will suffice. The two modes that I have illustrated in the workshops are unmediated strategies. The novice potter is *apprenticed*, here, in the sense that the emphasis is on the perfection of performances to the extent that only perfect products can be sold with the master potter's mark and only performances that are adjudged to be sufficiently close to this can be sold even as seconds. In the traditional apprenticeship, the passage to the next phase is marked by the production of the 'masterpiece'; an affirmation of competence, to be sure, but emphasis, here, is also on the object—the performance. I have used the term, *delegation*, to signify the strategy that I have illustrated in the scriptorium because this strategy seems to me also to characterise pedagogic delegation of responsibility, for example, in succession planning; for the transmitter to correct, rather than model, the acquirer's performances may be to inhibit the development of competence, which, after all, may ultimately, take a different form from that of the transmitter.

"The lower row of the schema is constituted by mediated transmission strategies and perhaps the most familiar is mediation in the school classroom. Here, most commonly, performances are of no lasting value in respect of the activity of *teaching*, they are there purely as indicators of competence. The use of aegrotat or compensatory assessment, where the significance of the performance indicator is modified in the light of contingent circumstances is a prime example of the *teaching* strategy. *Instruction* strategies are particularly prevalent in sets of instructions for particular performances that are generally unlikely to be repeated very often: instructions for the assembly of furniture delivered unassembled; emergency procedures or instructions for the use of the TV or telephone system in hotel rooms; instructions for adjusting your servo-powered seat in business class aircraft cabins. Of interest, here, is the tendency, in some circumstances, for schooling to distribute teaching and instruction to the most and least competent respectively (see Dowling (1990, 1998) in respect of school mathematics). Insofar as schooling constitutes mediated pedagogy, which is to say, the transmission of a recontextualised discourse (cf. Bernstein, 1990, Dowling, 1989, 1990, 1998, 2009, and Ernest 2006), this would tend to leave the least competent dependent on instruction in respect of a mythical practice. The tendency of schooling to recognise competence on the basis of social class[2] renders this all the more poignant.

"My use of the terms, transmitter and acquirer, may concern you, given your reflections on teaching and learning in your paper (Ernest 2006). However, I too reject the notion of simple transmission and acquisition. My approach understands pedagogic strategies and identities as being constituted contingently and in interaction (and its subsequent reflective recontextualisation), a view that owes its origins to interactive sociology (for example, Goffman, 1974, 1990; Strauss, 1997) and ethnomethodology (Garfinkel, 1967). The categories that I am introducing are strategies, not states, plays, not results. Their genealogies in specific contexts are not at issue here,[3] but, again, my schema enables the mapping in strategic terms of educational contexts at different levels of analysis. Again, the categories do not totalise any empirical setting—though the use of specific settings as illustrations may tend to make it appear that they do—nor does the schema itself totalise any setting, rather it provides an empirically derived, but logically exhaustive space for the announcement of strategies in and only in its own terms. This is a snap (for the time being), but one that may aid further description.

"When presenting the schema, transmitter strategies,' I am generally asked whether I've thought about acquirer strategies as well. In fact, I had reflected on this side of pedagogy before looking at transmitter strategies. In 1996 and 1997, Andrew Brown (you know him, my co-author in Brown & Dowling, 1998 and Dowling & Brown, 2010 amongst other collaborations) and I visited some high schools in the Western Cape area of South Africa.[4] We have presented an account and analysis of these visits in Dowling and Brown, 2009. On this walk I want to refer to a single finding. In two of the schools, students whom we interviewed were unanimous in their opinion that school knowledge had little or no value beyond the school. School success was important in giving access to higher education, which was, in turn, important in giving access to valued career opportunities. At a third school, however, we encountered a very different view, an astonishing view, perhaps; here is Andrew Brown speaking with students at the school (English is not the students' first language, possibly not their second or third):

> **AB:** ...is matriculation important for the careers that you have chosen.
> **P1:** Yes I think it is important because, to me I say the base of, if you got no matric you can't do anything because each, each, anything that I'm going to do, the base of it would be a matric. So I say it is important to have a matric and then going to do.... The base is matric and then take your career and then...
> **AB:** ...if I said that, OK, I've got some matriculation certificates here and I'll just write one for each of you and give them out.
> **P2:** ...it wouldn't count, you must have the base.

"School knowledge is here being constituted as vital in respect of the development of a competence that has value in respect of, well, all of the competences that are to follow and, in particular, to the most valued life opportunities relating to a 'career.' The irony of such earnestness is all the more apparent if I announce the nature of the schools. The first two were, respectively, an elite, primarily White school catering for the children of professionals and a dual medium (English and Afrikaans) school in a Coloured suburb, attended by children from a wide range of backgrounds, but including professionals. The third school was situated in an African informal settlement inhabited, primarily by casual labourers and school students, often living apart from their parents. The students speaking in the above extract were in a class for the 16–17 age group, but only the girls in this class were of this age; these students were men of between 22 and 33 years old who had returned to school after saving money to support themselves and, in one case, a wife and three children.

"Of course, all of the students have to acquire an appropriate level of competence in the practices that are their school subjects if they are to succeed, whether that success is understood as being based on certification or the knowledge itself. However, the strategies deployed differ between the first two and the third schools. The emphasis in the former settings is on what Bourdieu (1991) would refer to as symbolic capital, an objectified form, which, once acquired, may be 'exchanged' for other forms, in particular economic capital.[5] In the third school, however, what seems to be important is a to-be-embodied habitus.[6] This distinction in strategies is presented in the first of the central columns in the schema below (Figure 11.2).

"The final column of this schema distinguishes between strategies that focus not on the practice to be acquired, but on relations between participants. I do not have appropriate data to discuss these categories in relation to the schools, but I can draw on personal experience that I would imagine we share. So, I suggest that we both constitute certain relations as of value for themselves: close family members and friends, lovers, and so forth. The totality of relations of this kind places us at the hub of a radial sociogram; these relations are, in a sense, embodied as they materially impose on us. Less imposing relations are generally valued not in themselves, but rather for what they actually or potentially provide access to; in this sense, it is the objectified rather than embodied relations that are of value. The sociogram, here, is a network.

Culture	Acquirer Focus	
	Practice	Relations
Embodied	*habitus*	*hub*
Objectified	*symbols*	*network*

Figure 11.2 Acquirer strategies (From Dowling, 2009).

"This schema describes, but again in its own way, strategies that relate to Bourdieu's (1991) cultural and social capital that may be deployed by acquirers. As with the transmitter schema, I would expect individual acquirers and groups of acquirers to deploy more than one and potentially all of these strategies: students from the first two schools mentioned above must clearly be concerned with the development of habitus, even though they may consider the value of this embodied practice to be short-lived; students from the third school explicitly mention 'matric[ulation],' which is the symbolic form of the practice. Similarly, I have certainly relied on members of my hub (parents, partners, friends) to gain access to other relations and network relations have occasionally developed into embodied ones. As with the transmitter schema, the acquirer schema constitutes a logically complete, relational system that can be used to map specific pedagogic settings. They may also be used to identify confluences and antagonisms in such settings. Empirically, for example, doctoral research takes place substantially in the context of knowledge production and, as a doctoral thesis supervisor, I have often attempted to deploy *delegation*. My intention has been to provide an exemplar of, say, the analysis of empirical data in the hope that the student will be inspired to develop their own approaches that, whilst analytically competent, are not identical with my own. Sometimes this works: I like to think that it worked in the context of my own mentor (Basil Bernstein) and myself and indeed my own work now stands in (constructively) critical relationship to his. Sometimes, however, it does not work, because the student seems to understand my pedagogic action as *apprenticing*, limiting their own analytic activities to attempts to generalise my conceptual sketches that would, of course, have been based on a very restricted view of their data; they try to match performance rather than acquire competence. In other areas of doctoral work it may be appropriate to adopt *teaching* strategies. Attempts to develop competence in the use of statistical methods or in the interpretation of key theoretical works, for example. Then again, *instruction* is probably appropriate in respect of the formal production of the thesis document, entry to the examination and so forth.

"In mathematics education it seems to me that the optimistic introduction of 'investigations' into the school curriculum in the 1980s had the appearance of *delegation*. But can delegation ever be realised within the school? The rhetoric of investigations suggested that they involved the production of knowledge and that students were to act as mathematicians. Unfortunately, the only way that the knowledge that was produced could be legitimated was through its bureaucratic assessment against what were, in essence, behavioural objectives; this tended to push *delegation* to *instruction*.

"In a similar way, the acquirer schema might suggest that students focusing on *symbolic* acquisition may tend to view cynically teacher attempts to sell mathematics in terms of its potential use-value in diverse, non-school-mathematical contexts, thus emphasising *habitus* acquisition. My own analysis of school mathematics texts (for example, Dowling, 1998, 2007a) has revealed the torture that is generally (I think I would say always) necessary in the recontextualising of quotidian practiced in the formulation of what I refer to as the 'public domain' of school mathematics practice. The effects of such recontextualisation is the constitution of, for example, mythologised domestic practices that one would expect are easily recognised as such by students. If they are not recognised and if the students do buy into the public domain then they are duped by the lure of, shall we say, false habitus.

"The two schemas that I have introduced here have both been achieved via a flattening of an empirical chronotope; they are snaps that construct particular perspectives, just as a photograph constructs the camera position. However, the schemas also constitute a method for the construction of further analyses of settings other than their original empirical chronotopes as I have suggested here. Diachronic analyses or narratives are, of course, possible. Are such analyses also flat? In your three-dimensional analysis of number and counting you have emphasised both synchronic and diachronic axes. Perhaps I can now return to the question that I asked earlier: does this escape the flattening effect of the camera? Well, perhaps it does, but only in a fictional kind of way. A narrative is, after all, a completed text, it installs an inevitability in what is presented as an unpredictable walk. To borrow from Coleridge, I am able to suspend my disbelief in the implausibility of the narrative—which is to say its constructedness. In a good deal of academic analysis I am aided in this by the conventional elision of any representation of the self-awareness of the authorial voice—the method of the construction (I am not accusing you of this, by the way, though it would hardly be a pejorative observation were I to be doing so). Should I stray from the path that has been prepared for me, I am chivvied back on course by (what I refer to as, Dowling, 2009) traditional authority strategies, references to learned works that would assure me that any cynicism on my part would place me in an unsophisticated minority. In such works, the reticence of the self-awareness of the authorial voice renders narratives somewhat akin to Foucault's (1980) strategies without subjects. I want to put the subject back into the game. My subject begins with a very simple—I say, 'low discursive saturation' (DS⁻)—'internal language' that asserts, basically, that the sociocultural is to be understood as comprising autopoietic,[7] strategic action relating to the formation, maintenance and destabilising of alliances and oppositions, the visibility of which is emergent upon the totality of such action. My own analytic action then proceeds via the transaction of this internal language with empirical chronotopes. The outcomes of such transactions are, firstly, a highly complex, high discursive saturation (DS⁺) external language and, secondly, commentaries—snaps—of the chronotopes. That the external language consists largely of binary variables (and not, for example, digital or analogue spectra) is a consequence of the emphasis in my internal language or alliances and oppositions. The snaps are thus digitisings of the chronotopes. The articulation of DS⁻ internal language and DS⁺ external language constitutes my method. I have introduced elements of this method here in the two schemas that I have presented and in some of my other terminology. This terminology is generally associated with other schemas that are most fully represented in Dowling, 2009. In this work an organisational language of more than two hundred terms is established theoretically and empirically. Essentially, I guess, I boast about my cameras, not in any attempt to escape from their flattening effects—they are digital recontextualising machines the outputs of which are always further recontextualised in Photoshop—but in order to make explicit that which cannot be recovered from my snaps.[8] Where else might we point the camera?

"I mentioned earlier that the title of the early paper of mine that you cited was relevant to the theme of our walk; it's the word 'map' that is significant. I have often referred to the schemas such as the ones introduced here as maps, but they're not, are they. I've recently had my faith in maps shaken by a reminder about the navigation techniques of Pacific islanders provided by David Turnbull (2000). This, by the way, is a really wonderful book that I found very hard to put down and, on finishing it, I experienced the loss that I more usually associate with the finishing of good novels (the solace of Kermode's (1967/2000) 'sense of an ending' notwithstanding). Before saying more about Pacific navigation (and I don't intend to say very much here anyway) I want to offer an observation about the invisibility to the expert of embodied expertise—knowledge or skill. This is beautifully expressed in Wallace Stevens' poem, The Snow Man; I'll read it to you.[9]

> One must have a mind of winter
> To regard the frost and the boughs
> Of the pine-trees crusted with snow;
> And have been cold a long time
> To behold the junipers shagged with ice,
> The spruces rough in the distant glitter
> Of the January sun; and not to think
> Of any misery in the sound of the wind,
> In the sound of a few leaves,
> Which is the sound of the land
> Full of the same wind
> That is blowing in the same bare place
> For the listener, who listens in the snow,
> And, nothing himself, beholds
> Nothing that is not there and the nothing that is.
> (Stevens, 2001, p. 11)

"Need I say more; let's read it again, it's worth the diversion, I think, especially because of the contradiction between its setting and that of our philosophers' walk in the warm spring sun in which trees are not so much shagged with ice as frothed with blossom: does Stevens wipe the smiles from our faces or have we not yet been warm for long enough; just what is the nature of the walk from spring to winter and how is it to be navigated?

"Andrew Brown and I have recently been giving some thought to so-called professional conversion programmes (PCPs). These are adult education schemes that are intended to facilitate the re-skilling of people who have an expertise, the demand for which is currently on the wane, by training them in high-demand areas. The approach that has been generally adopted in the design of such programmes has been to attempt to identify generic skills; skills that trainees already have that will be useable in the target occupation and additional skills that they need to acquire. Conceptualised in this way, the walk from one occupation to another would seem to involve a simple process of packing up, before leaving, what will continue to be needed and collecting new baggage en route. The curriculum designer has the job of providing the road map, sketched out in terms of the generic skills required and (presumably) providing shortcuts where a particular trainee has already acquired a particular skill in the context of their original area of expertise. But then there's the problem of recontextualisation. We have three categories of participant here. Firstly, there is the trainee, who has expertise in a redundant activity. As a 'snow man' in that activity, he is not in a position to be explicit about his expertise. Secondly, we have the 'snow man' of the target expertise, who is in a similar, though healthier, position. Finally, we have the curriculum designer, who has expertise (we might hope) in curriculum design (though, given the snow man effect, not to the extent that s/he is able to render it in explicit form), but not in either the trainee's original practice or the target practice. Three viewpoints, but what do they look at?

"I was reminded of a diagram in Turnbull's chapter that he borrowed from Edwin Hutchins. I can't really show it to you here, while we're walking, but I can attempt to describe it. It seems that Micronesian navigators keep the star paths in their heads, wherever they go; this constitutes a kind of 'star compass.' When going for a 'walk' in the Pacific they identify a 'reference island' and establish the position of this *etak* in terms of the star compass at the beginning of the 'walk,' that is, they position it in relation to the rising and setting points of the stars in the compass. Of course, in 'walking' to a known destination, they also know the position of the *etak*—again in terms of the star compass—from there. The 'walk' is then conceptualised (and navigated) as the regression of the *etak* position in terms of the star compass; the navigator himself (sic) is considered, in this sense, to remain stationary. This seems to me (and, apparently, to Turnbull) to be a radical alternative to the 'god's eye view' (I think this is Donna Haraway's (1991) expression) of maps that seems to us to be such an obvious requirement for (or product of) almost any kind of walk, though de Certeau's (1984) *flaneur*—not really a navigator—is aso suggestive of this alternative. The Pacific navigator/*flaneur* strategy is also consistent with my preference for getting inside the game, describing transmitter and acquirer strategies rather than states of interaction. It is also not irrelevant that I seem to need to introduce god's eye maps to the introduction of my strategic schemas.

"I want to draw attention to the Turnbull/Hutchins description of the stationary navigator. Of course, the acting subject is always a 'snow man.' We might, then, consider the PCP trainee to be experiencing a changing horizon rather than as collecting generic skills and knowledges marked out on a map.[10] Would the provision of an *etak* help? What do we have in lieu of a star compass? At this point I have to confess that my thinking has not advanced very far in addressing the problem; possibly it's insoluble in a map-culture. However, we might, at least, focus our attention on the production of perspectives rather than movable skills. We do have reference points, in a sense: the practices at the point of original and target expertise. We also have three perspectives: that of the trainee, that of the practitioner in the target practice, and that of the curriculum designer; the latter is necessary as the motivator for the journey. Does this empirical chronotope provide us with the basis for the construction of a useable 'star compass'? This, of course, is an empirical question that I hope to be able to answer if and when Andrew and I complete our proposed research project. I would, though, very much appreciate your views and suggestions.

"Where does all of this leave us? Well, perhaps it might enable us to propose another set of alliances and oppositions in terms of analytic strategy. In that old paper of mine that you cited—'towards a conceptual map,' I think I wanted to establish exactly that, a map, a god's eye view, an objectivity. This certainly seems to characterise the products of my former mentor, Basil Bernstein. However, as I've tried to make clear during our walk this morning, I now try to get inside the game that I am watching to construct a subjectivity. Strategies, for me, have subjects that need to be identified in respect of their strategic action. This is not, of course, to claim that I avoid objectifying action—we are all snappers—but rather to suggest that I attempt a method that sees through the (objectified) eyes of the agents in my theorised chronotope. The 'maps' that may emerge from the accumulative development and deployment of my analytic schemas are analogous, perhaps, to the patterns of alliances and oppositions that are emergent upon agents' autopoietic action. In both my the earlier and later versions I am driven to make explicit the analytic apparatus that emerges from my transactions with the empirical chronotope, subsequently to be redeployed within it. In other words, I try to develop a DS⁺ external language. In your 2006 article you certainly render explicit an analytic apparatus in terms of your discussion of semiotic systems. However, I want to propose that this operates more at the level of an internal than an external language, that is, that this apparatus puts comparatively little pressure on the substantive analysis of the empirical chronotope. There is a sense, then, in which your approach is the opposite from mine: I have a weakly developed (DS⁻) internal language and a well-developed (DS⁺) external language; your internal language is DS⁺ and your external language DS⁻. I (Dowling, 2009) refer to your kind of configuration as a metaphoric apparatus; Bernstein also worked in this way. Your analysis, I think, fairly clearly objectifies the number and counting system (shall we say) from 'above,' which also marks it as distinct from my approach. I have established two analytic strategies, which is sufficient to enable me to present another schema; here it is (Figure 11.3).

"So, you are primarily a narrator and I am primarily a navigator. Clearly I am stretching the metaphorical use of the terms, somewhat, but they'll do for now. As I've mentioned before, these schemas are not intended to be totalising, so please don't feel that I'm trying to imprison you in a singular category and, anyway, my objectification of your analytic strategy serves purely as a pedagogic illustration. The flaneur strategy I have also referred to above. This strategy would seem to describe certain literary strategies, the strategy used by Geoffrey Hartman, for example, in his 'mildly deconstructive reading' (1987; p. 159) of three words ('a timely utterance') from Wordsworth's Ode (see also Dowling, 2005, 2009). Now the cartographers, who are they? Well, I suppose the modernist psychologists might fit in here, Freud (most obviously, 1976) of course, and also Piaget (let's use 1953) and Luria (1976). In each case, their well-developed internal language is augmented by an external language that is constituted by what Andrew and I (Brown & Dowling, 1998; Dowling & Brown, 2010) describe as 'elaborated description'; they argue their walks from their internal languages to their empirical chronotopes and their arguments comprise their DS⁺ external languages.

"Now, I wasn't going to mention this next schema, but I've already alluded to it and it articulates in a potentially interesting way with the previous one. It is constructed by crossing internal and external languages, again scaled in terms of DS. You'll recognize it because it was presented in

Perspective	External Language	
	DS⁺	DS⁻
Subjective	*navigator*	*flaneur*
Objective	*cartographer*	*narrator*

Figure 11.3 Strategies of analysis.

External Language	Internal Language	
	DS⁺	DS⁻
DS⁺	*metonymic apparatus*	*method*
DS⁻	*metaphoric apparatus*	*fiction*

Figure 11.4 Strategies of analysis.

a contribution to your journal (Dowling, 2007b, see also Dowling, 2009). Here it is (Figure 11.4).

"I think it should be clear why I've referred to your approach as broadly characteristic of a metaphoric apparatus and mine as a method; again, we seem to be in opposition, but only when viewed from these perspectives. I have also described the cartographers, Freud, Piaget and Luria in a way that renders their approaches with the metonymic apparatus strategy and possibly Hartman deploys something more akin to fiction.[11] This schema arose out of an engagement with Bernstein that is elaborated in Dowling, 2009; I think I'll not further discuss it this morning, if that's OK, but its interaction with the analysis strategies schema, with which it shares a dimension, might prove productive.

"The four schemas that I've introduced this morning are not maps. They are not really snaps either. If I'm going to pick a metaphor, then I guess I'd have to say that they represent different configurations of my cameras; they are used to produce snaps. My introduction is intended as a kind of users' manual. As a navigator, my intention is to try to make my navigational apparatus as visible, as explicit as possible. Now I've done most the talking on our walk, but I think I can hear the beginnings of a protest. On the one hand, I seem to be privileging an approach that makes its method explicit but, on the other, I have asserted—with the help of Wallace Stevens—that genuine expertise cannot speak about itself. However, perhaps the camera metaphor helps here if we compare the photographer with the (recontextualised) watercolourist: the photographer deploys a highly complex apparatus, the watercolourist a simple one. This says nothing at all about the relative quality of the images they respectively produce. My analytic action is analogous not simply to the photographer, but to photographer and camera designer. My approach has the advantage of producing not only comentaries on the empirical chronotope, but apparatus cabable of being recruited in the production of more commentaries by other analysts. In other words, my approach incorporates a form of pedagogy.

"Many of our companions on our walk are using the camera facility of mobile phones. I suspect that some of them will transmit their snaps using their phones' email facilities and some of these will be saying something like, 'this is how it is here,' eliding the recontextualising and flattening effects of both camera and photographer. Some cartographers and navigators do likewise. The grounded theorists, for example (perhaps especially Barney Glaser, 1992) have provided us with very sophisticated apparatuses for the analysis of qualitative data the correct use of which, we are told, will enable theory to emerge from the empirical chronotope. They are, though, suspiciously silent on the nature of the subject that is to wield this apparatus. The critical discourse analysts—here I'm thinking of Norman Fairclough (1995; Chouliaraki & Fairclough, 1999) again present a highly sophisticated apparatus, which is essentially linguistic, but they really want to bring us truths about the effects of power in society. Language and power are mediated, in their schemes, by the categories discourse and genre, I recall this from Fairclough:

> There are no definitive lists of genres, discourses, or any of the other categories I have distinguished for analysts to refer to, and no automatic procedures for deciding what genres etc. are operative in a given text. Intertextual analysis is an interpretive art which depends upon the analyst's judgement and experience. (Fairclough, 1995; p. 77)

"Well, I would be perfectly happy about this were it not for my suspicion that they had made up their minds on what they wanted to say about power before they opened their linguistic camera bag. Maybe I do them an injustice, but I do wonder if they are ever surprised by their own commentaries; perhaps I'm deceiving myself: perhaps none of us is every really surprised.

"So, I'm a camera toting snow man (hope I don't melt in the sun) who likes to boast about his cameras and explain their workings to anyone who'll listen. I think Nishida would have approved of the openness of this position, so wouldn't object to us trespassing on his path. I'm still a little worried, though, about Gide's remark; has this all been said before? Well, I'm rather attracted to Foucault's comment on commentary (all of this, after all, is a commentary on something—the philosopher's walk, perhaps):[12]

> Commentary's role . . . is to say at last what was silently articulated 'beyond' the text. By a paradox which it always displaces but which it never escapes, the commentary must say for the first time, what had, nonetheless, already been said, and must tirelessly repeat what had, however, never been said. (Foucault, 1981, p. 58)

Perhaps all of this has been said before, but it hasn't been said here.

"I understand it's your birthday soon: お誕生日おめでとうございます. Thanks for joining me this morning; please feel free to borrow one of the cameras anytime you think it might be useful; I'd be happy for you to disassemble and reassemble it, make some improvements, perhaps. You know, it really is a beautiful day for a walk. But then it almost always is, don't you find."

NOTES

1. See http://www.quotegarden.com/experience.html.
2. I'm afraid you credited me with having published this work a little in advance of its actual publication; it was originally published in a microfiche journal in 1989 and subsequently in a collection in 1991, not in 1988 ;-).
3. See, for example, Bernstein, 1977; Bourdieu & Passeron, 1977; Dowling, 1991b, 1991c, 1998; Sharp & Green, 1975; the work of the seventies in this exemplary list is dated, but should not be forgotten.
4. But see Hunter (1994) for an interesting genealogy of the school.
5. These visits were made possible by an Overseas Development Agency funded link between Brown and myself and the universities of Cape Town and the Western Cape.
6. I have always had a problem with Bourdieu's metaphor of capital exchange; after all, it is only economic capital that actually circulates (see Dowling, 2009).
7. I borrow this term from Maturana & Varela (1991), though with a less developed interpretation.
8. I am only an amateur and not very skilled photographer, but it seems to me that many camera features and settings—such as the focal length of the lens, focusing point, aperture, shutter speed, white balance setting and so forth—cannot be inferred from the photograph alone without further knowledge or assumptions about the setting within which the snap was taken. Inference of other features—lens quality, perhaps—might be made primarily on the basis of a knowledge of cameras.
9. I was introduced to Stevens' poetry by Soh-young Chung, who uses poetry—including this poem—methodologically in her thesis (in preparation) on the sociology of literary studies.
10. Given that the skills and knowledges are constructed by individuals other than s/he, perhaps a computer game avatar searching out concealed ammo and medpacks on their murderous adventure might provide a better metaphor for the traditional programme.
11. Not all literary criticism is like this. Louis Montrose (1989), for example and others that Chung (2009) describes as "theorists" are more appropriately described as deploying metaphoric apparatuses.
12. I was reminded of this extract by Soh-young Chung.

THOSE ENCOUNTERED ON THE WALK
(SOME UNDERFOOT, BUT NOT TRAMPLED)

Bernstein, B. (1977). *Class, codes and control: Towards a theory of educational transmissions*. London: RKP.

Bernstein, B. (1990). *Class, codes and control volume IV: The structuring of pedagogic discourse*. London: RKP.

Bloom, H. (1973). *The anxiety of influence: A theory of poetry*. New York: Oxford University Press.

Bourdieu, P. (1991). *Language and symbolic power.* Cambridge: Polity Press.

Bourdieu, P., & Passeron, J-C. (1977). *Reproduction in education, society and culture.* London: Sage.

Brown, A. J., & Dowling, P. C. (1998). *Doing research/reading research: A mode of interrogation for education.* London: Falmer Press.

Chouliaraki, L., & Fairclough, N. (1999). *Discourse in late modernity: Rethinking critical discourse analysis.* Edinburgh: Edinburgh University Press.

Chung, S-y. (2009). *The crafting of crisis: A sociological analysis of the "Cultural Studies Paradigm Shift" in literary studies.* PhD Thesis. Institute of Education, University of London.

Cohen-Mushlin, A. (2008). *"A School for Scribes."* Presented at Comité International de Paléographie Latine XVIth Colloquium: Teaching Writing, Learning to Write. University of London. September 2–5, 2008.

de Certeau, M. (1984). *The Practice of Everyday Life.* Berkeley: University of California Press.

Dowling, P. C. (1989). "The contextualising of mathematics: Towards a theoretical map." *Collected Original Resources in Education 13*(2).

Dowling, P. C. (1990). The Shogun's and other curricular voices. In P. C. Dowling & R. Noss (Eds). *Mathematics versus the national curriculum.* Basingstoke: Falmer.

Dowling, P. C. (1991a). The contextualising of mathematics: Towards a theoretical map. In M. Harris (Ed). *Schools, mathematics and work.* London: Falmer.

Dowling, P. C. (1991b). "Gender, class and subjectivity in mathematics: a critique of Humpty Dumpty." *For the Learning of Mathematics. 11*(1), 2–8.

Dowling, P. C. (1991c). A touch of class: Ability, social class and intertext in SMP 11–16. In D. Pimm & E. Love (Ed.), *Teaching and learning school mathematics.* London: Hodder & Stoughton.

Dowling, P. C. (1998). *The sociology of mathematics education: Mathematical myths/pedagogic texts.* London: Falmer.

Dowling, P. C. (2005). *A timely utterance.* European Systemic Functional Linguistics Conference and Workshop. King's College, London. Available at http://homepage.mac.com/paulcdowling/ioe/publications/dowling2005/timely_utterance/index.htm and also in Dowling, 2009.

Dowling, P. C. (2007a). Quixote's science: Public heresy/private apostasy. In B. Atweh et al (Eds). *Internationalisation and globalisation in mathematics and science education.* Dordrecht: Springer.

Dowling, P. C. (2007b). Social Organising. *Philosophy of Mathematics Education Journal 21*, 1–27.

Dowling, P. C. (2008). *Mathematics, myth and method: The problem of alliteration.* Presented at Kings' College, London, 25th November 2008. Available at http://homepage.mac.com/paulcdowling/ioe/publications/dowling2008a.pdf

Dowling, P. C. (2009). *Sociology as method: Departures from the forensics of culture, text and knowledge.* Rotterdam: Sense.

Dowling, P. C., & Brown, A. J. (2009). "Pedagogy and Community in Three South African Schools: An iterative description." In P. C. Dowling (Ed.), *Sociology as method: Departures from the forensics of culture, text and knowledge.* Rotterdam: Sense.

Dowling, P. C., & Brown, A. J. (2010). *Doing research/reading research: Re-interrogating education.* London: Routledge.

Ernest, P. (2006). "A Semiotic Perspective of Mathematical Activity: The case of number." *Educational Studies in Mathematics 61,* 67–101.

Fairclough, N. (1995). *Media discourse.* London: Edward Arnold.

Foucault, M. (1980). *Power/knowledge.* Brighton: Harvester.

Foucault, M. (1981). "The order of discourse." In R. Young (Ed). *Untying the text: A poststructuralist reader.* London: RKP.

Freud, S. (1976). *The interpretation of dreams.* London: Penguin.

Garfinkel, H. (1967). *Studies in ethnomethodology.* Englewood Cliffs: Prentice-Hall.

Glaser, B. G. (1992). *Basics of grounded theory analysis: Emergence versus forcing.* Mill Valley: Sociology Press.

Goffman, E. (1974). *Frame analysis.* New York: Harper.

Goffman, E. (1990). *The presentation of self in everyday life.* Harmondsworth: Penguin.

Haraway, D. J. (1991). "A cyborg manifesto: Science, technology, and socialist-feminism in the late twentieth century." In D. J. Haraway (Ed.), *Simians, cyborgs and women: The reinvention of nature.* London: Free Association Books.

Hartman, G. (1987). *The unremarkable Wordsworth.* London: Methuen.

Hunter, I. (1994). *Rethinking the school: subjectivity, bureaucracy, criticism.* St Leonards: Allen & Unwin.

Kermode, F. (1967/2000). *The sense of an ending: Studies in the theory of fiction.* NY: Oxford Univesity Press.

Lave, J., & Wenger, E. (1991). *Situated learning: Legitimate peripheral participation.* Cambridge: CUP.

Luria, A. R. (1976). *Cognitive development: Its cultural and social foundations.* Cambridge, MA: Harvard University Press.

Maturana, H., & Varela, F. (1991). *Autopoiesis and Cognition: The realization of the living.* NY: Springer.

Mauss, M. (1979). *Sociology and psychology: Essays.* London: RKP.

Montrose, L. A. (1989). Professing the Renaissance: the poetics and politics of culture. In H. A. Veeser (Ed.), *The new historicism,* (pp. 15–36). New York: Routledge.

Piaget, J. (1953). *The child's conception of number.* NY: Humanities Press,.

Sharp, R., & Green, A. (1975). *Education and social control.* London: RKP.

Singleton, J. (1989). "Japanese folkcraft pottery apprenticeship: Cultural patterns of an educational institution." In M. Coy (Ed), *Apprenticeship: From theory to method and back again.* Albany: State University of New York Press.

Stevens, W. (2001). *Harmonium.* London: Faber & Faber.

Strauss, A. (1997). *Mirrors and masks: The search for identity.* London: Transaction.

Turnbull, D. (2000). *Masons, tricksters and cartographers.* London: Routledge.

CHAPTER 12

GEOMETRY

Tales of Elegance and Love

Tim Rowland
University of Cambridge, UK

INTRODUCTION

Paul Ernest and I were appointed to mathematics education posts in Cambridge at the same time—a time closer to the beginning of our careers than to our retirements. I enjoyed working with Paul, and we seemed to have a great deal of time in those days just to sit and talk with each other and our colleagues at Homerton College. I have followed Paul's illustrious career with interest and pleasure ever since. This is not the place to assess his contribution to our field, and I am not the person to do so. Suffice to say that my inspiration for writing this particular offering is Paul's generous and consistent appreciation of the wide range of ideas and scholarly outputs of his colleagues, and his ability to enthuse about and affirm them. I was tempted at first to channel my efforts for this chapter into a short piece on mathematical logic—one of our shared interests all those years ago in Cambridge. But then I also knew that I had something else simmering away in my head: not logic, as such, and not really mathematics education, but I

Relatively and Philosophically E^{a}rnest, pages 191–213

was excited by it, and writing it up as my contribution here is appropriate, if somewhat different from the alternatives that I had considered.

The geometry problem at the heart of this chapter came to my attention at a seminar given by Orit Zaslavsky in Cambridge early in 2006. The problem features in Zaslavsky (2005), a paper about the design and implementation of mathematical tasks that evoke uncertainty for the learner. Zaslavsky presents it in the form shown in Figure 12.1, although the original textbook problem is "Prove that $\alpha = 60°$." The point is that not revealing [the measure of] the angle α provoked a debate between students to resolve their competing claims. One student, Bob, claims that $\alpha = 60$ (angles will be in degrees throughout), and presents a proof that he found in a textbook, based on what looks like an ingenious construction (Figure 12.2). However, this fails to convince Ruth, another student, who believed at first that $\alpha = 55$. Orit goes on to write about ways in which this task was developed over time to provoke uncertainty and debate. These include inviting some of the students to prove or refute that $\alpha = 60$, and the rest to do the same

Given that ABCD is a square, what is the measure of the angle α?

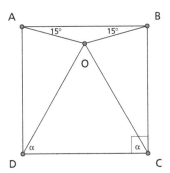

Figure 12.1

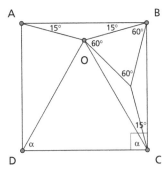

Figure 12.2

with $\alpha = 55$. Her point is that this doubt-provoking task fostered productive thinking about a wide range of issues central to mathematics, such as validity and certainty, multiple proof methods, existence and uniqueness, examples and counterexamples.

Now, at the seminar, Orit was unable to devote much time to the details of the students' responses to the task, although her paper (Zaslavsky, 2005) includes two teachers' proofs by contradiction that $\alpha \neq 55$, including one that would show that $\alpha \neq \theta$ for every $\theta \neq 60$.[1] Anyway, I became curious about how to prove that $\alpha = 60$ during the seminar, and the feeling wouldn't go away. Orit did show us the student Bob's diagram (Figure 12.2) but, I didn't find it very helpful. Moreover, I had the "I'd never have thought of that" reaction to it, which was quite alienating.

TONY P.

Later the same week, I "phoned a friend." Well, e-mailed. Tony Pay read mathematics at Cambridge a long time ago, but has earned his living as a musician. I wrote:[2]

> OK, here's a little "math" in case you find yourself without a crossword to do. ABCD is a rectangle with AB = 2BC. P is a point on AB such that angle PDA = 15 degrees. The original problem says "prove that angle PCB = ** degrees," but I'll say "what is angle PCB?" I then ask that you prove your assertion by Euclidean methods (i.e., methods known to Euclid, the stuff of Durell). By the way, I bought a copy of Durell from Abe[3] about 5 years ago.

The reference here is to Clement V. Durell (the "V" is for Vavasor...), whose textbooks had circumscribed, in a benign way, our East London Grammar School mathematical education between the ages of 11 and 15. Durell, the son of the Rector of Fulbourn, in Cambridgeshire, was a Cambridge wrangler[4] who spent most of his career teaching at Winchester school. He was a prolific and very successful writer. When he died in 1968, his estate was worth around five million pounds in today's values. His school geometry text (Durell, 1939), was actually rather good, and I think that Euclid himself would not have been disappointed with it. A certain Carl E. Linderholm rather unkindly dedicates his 1971 book *Mathematics Made Difficult* to "Clement V. Durell, M.A., without whom this book would not have been necessary."

I note here that I had reframed the original problem (Figure 12.3). It was easier to describe this version in words, and my efforts to solve the original had convinced me that it would be sufficient to work with just one half of the original line-symmetrical figure. In a way, I now regret imposing that restriction on others.

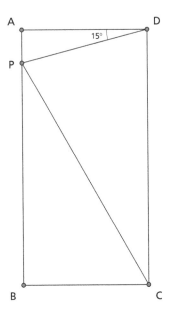

Figure 12.3

Tony replied:

What was Durell called, exactly? I might try to do the same. What surprised me was that I didn't know this result. I suppose the "right" way to do it is to prove directly somehow that if AB = 2, BC = 1, and PDA = 15 degrees, then PC = 2. What I did was: AP = tan 15 = t say, and then sin 30 = 1/2 = 2t/(1 + t squared), solve quadratic for t giving AP = 2–sqrt 3, from which PB = sqrt 3 and so triangle PBC is half an equilateral triangle, and the angle is 60 degrees. It's not very nice, but I can't think of a natural way to use the 15 degrees in a construction.

I replied with the name of Durell's book, adding:

Do you also remember (for the Cambridge Schols[5]??) his *Modern Geometry* (1st Ed 1920)—very elegant. I have that too—must have stolen it from somewhere... I have since found that connoisseurs of Eucl Geom drool over L. Roth—*Modern Elementary Geometry*, Nelson, 1948..." such elegant riders...").

The reference to Roth and to connoisseurs was prompted by an enjoyable experience just a few years earlier working on a government-funded secondary algebra and geometry project with Tony Barnard and a small group of talented teachers. Our contribution was to devise and trial materials for teaching Euclidean geometry. A guiding principle behind the "scheme"—entirely novel to two younger members of the team—was that the "starting points" were made explicit, and absolute rigour was demanded in the deductive arguments based on these axioms, although that makes it sound more scary than it was. The enthusiasm of this team was invigorating and highly infectious. Incidentally, Leonard Roth, another Cambridge wrangler, was born in Edmonton around 1905. He became a Reader at Imperial College and was only 64 when he died in a car accident.

I added, [> It's not very nice] Well, it is in a way, but goes beyond Durell New Geom. My rough and ready effort was to try to show that if tan15 + tan(a) = 2 then a = 60, but I can't do it nicely."

Tony (Pay)'s reply:

> I get it. What clears it up is to look at it the other way around. Imagine a general rectangle ABCD with AB = 2, BC = y, say. Now *construct* CP = 2. It's not difficult to show angle PDA = 1/2 BPC. Then, what more natural for the examiner to choose y = 1, PDA = 15 degrees, and set the problem backwards. And in fact, all that can be done (backwards) just using Durell/Euclid :-)

> Thanks for telling me this — it was fun to think about! And I've ordered my Durell . . .

As I read this, I thought that the observation that if CP = CD, then, *irrespective* of the dimensions of the rectangle, ∠BPC = 2∠PDA—easily proved by angle-chasing—was a significant and pleasing insight. In the case that PDA = 15, ∠BCP would have to be 60, and BC = 2 in turn. However, I replied:

> This is very neat, but I'm still stuck! I see that if CP = CD and CD = 2AD then it follows that ADP = 15–the crucial insight being your angle PDA = 1/2 BPC. But (and I hope this isn't being pedantic) I can't deduce CP = CD from the premises: ADP = 15 and CD = 2AD.

There were further exchanges before the conversation fizzled out. Tony was working hard to find a "natural" context for the result, one of them very much in the spirit of dynamic geometry, with P sliding up and down AB, with corresponding consequences for the rest of the diagram. It was interesting for me to see that Tony was quite fired up by this problem, nearly 40 years after having—supposedly—"given up" mathematics. But I'll move on.

TONY B

By now, I was curious to know what the other Tony—Barnard—would make of the geometry problem. Tony is a mathematician, with an active interest in mathematics education—a kind of complement, or mirror image, of myself, in fact. I sent this Tony the problem, and the gist of my discussions with the other Tony. He replied:

> Dear Tim, This nice problem is similar to one we set in our annual problems drive at King's [London] many years ago. I have to admit that I'd forgotten about it until reminded by my friend John who has three nice solutions. The following is adapted from one of them and I've tried to make it as elementary as possible (in the sense that it uses the minimum amount of theory). In the description below, numbers will be degrees, but the word 'degrees' will be omitted for brevity.
>
> Consider the rectangle ABCD as the lower half of a square DEFC. The triangles PDA and PEA are congruent (SAS), so angle AEP = 15 and angle FEP = 75.

Given that I had posed my email version of the problem by reference to half of the original square, it was interesting to see that Tony had now reinstated the square.

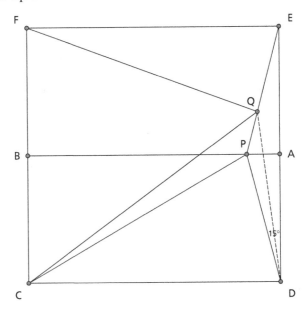

Figure 12.4

> Now let Q be the point on EP (or EP produced) for which FQ = FE. Then angle FEQ = 75 (base angles of an isosceles triangle) and so angle EFQ = 30. Therefore angle QFC = 60. It now follows that triangle CFQ is equilateral (as FQ = side-length of the square DEFC). Therefore angle FCQ = 60 (hence angle DCQ = 30) and also CQ = CD (side-length of the square again). So angles CDQ and CQD are (also) the base angles of an isosceles triangle whose other

angle is 30. Thus angle CDQ = 75. Hence angle PDQ = 0. This means that P and Q must be the same point. Therefore angle PCB = angle QCB, which is equal to 60.

I recognised the care with which Tony justified each and every step of the deductive argument, even down to "Hence angle PDQ = 0," when it would be natural to say "But CDP also equals 75, so P and Q must be the same point." I was also intrigued by a strategy that proposes a point (Q here) that has certain desirable properties, and constructs an argument that inexorably identifies this point with one "already" in the diagram. There was more:

P.S. The old King's problem (but we don't claim to have invented it!) was: ABCD is a square, and E is the internal point such that angles EAD, EDA are both 15 degrees. Show that EBC is an equilateral triangle.

Indeed! Figure 12.1, with the point O now re-labelled as E.

The problem you sent me is converted into this by reflecting everything in AB (the AB of the problem you sent, that is). John's solutions of the 'King's' problem are as follows.

1. Reflect E in BD to give F, and show DEF is equilateral. It then follows that CFD, CFE are congruent, so that CE = CD = BC; and similarly for BE.

Ah—Figure 12.2: Bob's ingenious construction. OK. The detail of this solution is omitted, but I can see how I could fill them in if required. Required by whom?

2. Let the circle centre B through A meet AE in E': then by the alternate segment theorem, AE' subtends an angle of 15 degrees, so that ABE' is 30 and E'BC is 60, that is, E'BC is an equilateral triangle. By symmetry, E' = E.

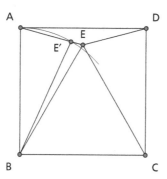

Figure 12.5

That one needs some unpacking, and eventually I arrive at the following. AD is a tangent to the circle through A and E′, centre B. By the alternate segment theorem (and the 'angle at the centre' theorem), $\angle ABE' = 2\angle E'AD$. But A, E and E′ are collinear, so $\angle E'AD = \angle EAD = 15$, $\angle ABE' = 30$ and $\angle CBE' = 60$. Since BC($=$ BA) $=$ BE′, triangle CBE′ is equilateral. The "by symmetry" seems to be saying that if E″ is a point on ED defined in the same way as E′, but with circle centre C, then both BE′C and BE″C would be equilateral triangles, "and this could only happen if E′ and E″ were the same point, at E." Specifically—$\angle E'CE'' = 60{-}60 = 0$. Given that E′ and E″ are defined to be points on AE and DE respectively, $\angle E'CE'' = 0$ when both points coincide at the intersection of the two lines i.e., at E. Again, I like the strategy of defining a point one way and then constructing an argument to identify it with another point. Well, not another, of course, in the end.[6]

> 3. Let P′ be the reflection of P in DA, and let A′ be the reflection of A in DP. Then angle ADA′ is 30, and a rotation of 30 about D sends P′ to P and A to A′, and therefore sends the line P′A (that is, AB) to the line PA′. By the rotation, angle A′PB is 30. Let the line PA′ meet DC at C′. Then DA perp AB, so DA′ perp PC′; also angle A′DC′ = 90–30 = 60, and triangle DA′C′ is a 30–60–90 triangle. Further, DA′ = DA (by the rotation), so DC′ = 2DA′ = 2DA, whence C′ = C, and PCB = complement of A′PB = 60.

A little later, Tony helpfully remarked, "As you must have noticed, the third of John's solutions was directly for the one you sent, rather than for the 'King's' problem."

After several false starts, I had noticed! Once again, I noted the construction of the point C′, and its subsequent identification with C. (I think it's time to give that strategy a name: perhaps the "identical points" strategy). I was a little surprised that Tony had forwarded a proof couched in the language of geometrical transformations, though it worked for me without complication, despite the by-now familiar "How did they think of that?" experience.

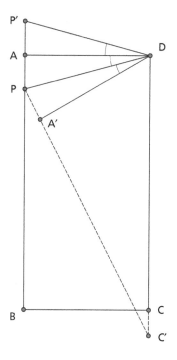

Figure 12.6

Tony's most recent message continued:

Looking at it more closely, I see that this doesn't really need any properties of transformations. All you have to do is the following.

Let X be the point on DC (or DC produced) for which angle BPX = 30. Then angle DPX = 75 (angles on a straight line, because angle DPA = 75). Now let Q be the point on PX such that angle PDQ = 15. Then angle DQP = 90. Therefore triangles ADP and QDP are congruent (AAS). But DX = 2DQ. (This follows either from the fact that $\sin(30) = \frac{1}{2}$ or by easily showing that, if R is the point on DQ produced for which QR = DQ, then triangle DRX is equilateral.) Therefore DX = 2AD = DC. Hence X = C and so angle CPB = angle XPB = 30. Therefore angle BCP = 60.

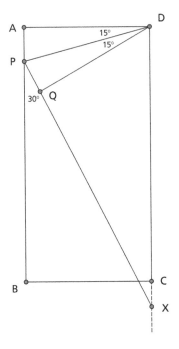

Figure 12.7

Yes, a very neat transformation of the "transformations" proof. What would Tony the musician think of it? His reply:

> How did we miss *that*?? Though I thought the proof would have been better with a variant P (P' say) to make the *length* right, as I suggested before [the dynamic geometry-like proof]:

> Let X be the point on CB produced with CB = BX, and P' the point on AB such that CP' = CD = CX. Then CP'X is equilateral, DCP' = 30, P'DA = 15. PDA = 15. Hence P = P'.

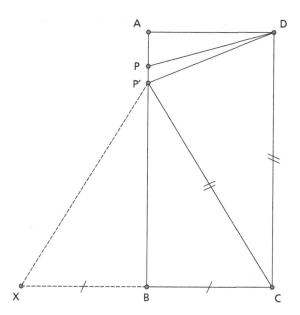

Figure 12.8

(no doubt the other Tony would insert a statement saying that ∠PDP′ = 0)

...but that again seems to require a knowledge of the answer, which the construction of X in 'solution 1' [the proof that I'd sent] doesn't really. (After all, you've got to get the 15 degrees involved somehow, and sticking a 60 degrees in to leave another 15 in the right angle is something you might think of.)

Off to OZ, NZ, Kuala Lumpur and Hong Kong...

...and a little later—:

But before I go, I just wanted to say that the 20–20 hindsight, motivated proof number 0 is:

Rectangle ABCD, AB = 2BC, P on AB s.t. angle PDA = 15 degrees, what is angle PCB?

How do we use AB = 2BC? ... Construct perpendicular bisector of DC through midpoint X, say.

How do we use angle PDA = 15 degrees? ... Construct line through D at 15 degrees to DC meeting that perpendicular bisector at Y, say. Now join CY, CP.

This creates three congruent triangles DPA, DYX, CYX, so triangle DPY is isosceles with vertex 60 degrees, so equilateral. So DY = PY = CY, triangles YDC, YPC are congruent, PCD = 30 degrees, and PCB = 60 degrees.

That's what I was trying to say all along!:-) Tony

Sketching the diagram, as I hope you will, shows how Tony's efforts—with hindsight admittedly—use and build on the information given in "Bob's ingenious proof." It remains ingenious, but my earlier sense of alienation has now gone.

REX

A week later I emailed the problem to my (former) colleague, Rex Watson. Rex frequently sends me problems (inventions of his fertile and evidently restless mathematical intellect) and I regret having less and less time to dwell on them. I had slightly mis-remembered my earlier notation: now the rectangle ABCD has BC = 2, AB = 1, P on BC such that BAP = 15. Rex's solution came quite quickly:

> Let M be the midpt of AP and let the perp bisector of AP meet AD (produced if necc) at Q. Let R be on BC s.t. QR is perp to AQ.
>
> Now AQM = 15 (easy) and QMP, QMA are congruent (SAS), so AQP = 30 and so PQR = 60.
>
> But QR = 1 so PQ = 2 (by the usual equilateral triangle argument). Therefore QA = 2 i.e. Q is D, so PDC = 60.

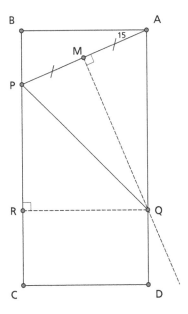

Figure 12.9

Another "identical points" proof, but a new one. I forwarded it to Tony B, with the comment, "I think this is different from the others... and there's something very direct about it."

Tony's reply:

> Thanks for sending this. It's a lovely solution, but why bother with the congruent triangles? Why not just take Q on AD such that APQ = 75? The 30–60–90 triangle PQR forces PQ to be 2, and so Q to be D. This solution is equally direct in its alternative form (where you have a 15–15–150 triangle inside a square and have to show that the "other" one is equilateral).

A characteristic of both Tonys is to appreciate a new solution, but to see how it might be refined and improved. Each solution is a kind of launch pad for yet another.

DEREK

Four months have elapsed, and now it's July. I receive an email from Tony P.

> I was writing to Del Smith, and thought to send him "the problem" in the form, ABCD square, P point inside s.t. angle PAD = angle PDA = 15, show PBC equilateral eschewing trig.

Interestingly, Tony has converted my rectangle form of the problem back to the original. I should have left it alone. Derek ("Del" to we East Londoners) Smith was also at school with us. He played a mean jazz piano, read mathematics at the fledgling Sussex University,[7] and Tony had evidently kept in touch. Del was now retired in Cyprus, having taught in a school there for some years. I met up with him during a trip to Cyprus the following year: once we had accounted for the 40 years since we had last seen each other, Del was keen to expound his latest ideas about relativity.

Tony passed on Del's proof:

> Construct circumcircle of ADP, centre O, rad = r
> Reflex $\angle AOD = 2^\wedge APD = 300$, $\angle AOD = 60$
> OD = OA = r, AD = r, ABCD square, CD = r
> DC//OP, DC = OP, OPCD //gram and CP = OD = r
> PBC equilateral
> Have you got a neater one?

I will confess now to a liking for proofs that introduce circles and circle theorems when there were none in the problem as stated. The statement of this one is somewhat minimalist. In particular, from the fact that

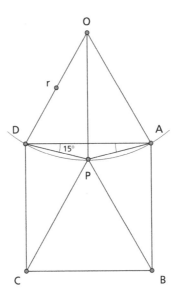

Figure 12.10

∠AOD = 60 and AB = OA (both radii of the circle) we can say that triangle OAD is equilateral, and so the radius of the circle is the same as the side of the square. I find that very elegant, and surprising. Strictly speaking, there is some work to be done to show that OP is parallel to DC, and it could be done, but the brevity of the proof is something of an illusion.

I don't seem to have referred Del's proof back to Tony Barnard (but it's never too late...)

VESNA

Now fast forward nearly three years. The year is 2009. We have a good group of mathematics education masters students this year, among them Vesna Kadelburg—another Cambridge wrangler. Vesna completed her mathematics PhD not long ago, and her day-job is teaching in a local school. She is also the deputy leader of the UK team for the International Mathematics Olympiad. At a recent supervision, it occurs to me that she might enjoy The Problem, so I jot it at the end of my comments on one of her essays, and she hurries away to a Research Methods session. Now I should explain that one of the readings for our masters course is my account (Rowland, 2003) of solving a problem—one that I learned from Tony Pay, in fact—while walking along the river from Ely to Cambridge. In that paper, I argue that it mattered where I was and what I was doing: that the network of ideas and memories which we assemble in the course of solving a particular problem has temporal, spatial and emotional components. Vesna's narrative account is a deliberate and playful imitation of the genre!

> Dear Tim, Here is my story about solving the geometry problem. I don't have your original diagram any more, so sorry if the labelling is not the same. [It's similar but not quite the same]. ABCD is a rectangle with BC = 2AB. P is a point on AD such that ∠ABP = 15. Prove that ∠PCD = 60.

> I started by thinking: if the claim is true, what else needs to be true? If ∠PCD = 60, then PCD is a 30–60–90 triangle, which I spotted straight away, because it is my favourite triangle as it comes useful in many different contexts (from evaluating trig ratios of special angles to solving some pretty obscure geometry problems). I know that this is half of an equilateral triangle, hence PC = 2CD, so PC = BC. Thus if I can show that PC = BC, then I'm done.

> I can see two ways of doing this: Pythagoras or trigonometry. First, label some sides: AB = a, BC = 2a, AP = x so that PD = 2a–x. I can play with Pythagoras on triangles ABP and PDC. As I know what I'm trying to get, I should be able to fudge the algebra. I try this for about two minutes, and then give up . . .

Would anyone like to try this idea for three minutes, just in case?

> . . . and decide to launch into some trigonometry. I started this towards the end of the break in a [research methods] lecture, then had a short interruption to complete a set task, and now have a few minutes spare before the next bit of the lecture starts.

> I decide to go with the trigonometry. I can find by using double-angle formula that cos15 = (1 + root(3))/2root(2) and cos75 = (root(3)–1)/2root(2). I therefore write BP = acos15 [presumably a/cos15 intended] and then use cosine rule in triangle BCP to find that indeed PC^2 = 4a^2, so PC = 2a, as required. The algebra is not too bad, especially as I know what I'm aiming for. This takes about 5 minutes.

Well, it looks good to me, though it took me more than 5 minutes to check the details. And yes, it is trigonometry. But there is more:

> I then get back to the lecture, but keep thinking that I'd now like to find a nice solution to the problem. As I want CP = CB, I start scribbling some circles, hoping to use circle theorems about angles.

So Vesna likes to use circles too.

> I try a circle with centre C passing through B (and so hopefully P), but don't see anything immediately. The lecture is over and I get on my bike. It's good that I have solved the problem, but I want a nicer solution.

That elusive "nice" quality . . . elegance. The aesthetics and the affective response to it are very much to the fore in this contribution, and that of the two Tonys. Read on.

You wouldn't set me a problem which is plane trigonometry (although it is a nice example of how powerful trigonometry actually is, even if angles don't at first appear to be "nice"!).

I said something similar about Tony Pay's problem in Rowland (2003). Knowing the problem poser sets up certain expectations of the problem itself.

I am cycling and still thinking about circles, when it hits me that I have solved the problem long time ago, I was just not using my logic in the right direction. In order to avoid accidents, I try thinking about something else until I get home, and then immediately get a piece of paper to get my thoughts down in writing (I like to see the flow of logic in writing, that's the way I check that it does actually work). I immediately fall in love with this problem—the geometry is trivial, but the logical construction is beautiful!

Any commentary from me must surely be superfluous.

My solution: Let Q be a point on AD such that <DCQ = 60. There is clearly only one such point, as it is obtained by drawing a line through C making a 60-degree angle with CD, and intersecting it with AD. Then CDQ is a 30–60–90 triangle, hence CQ = 2a. But then CQB is isosceles, and ∠QCB = 30, so ∠QBC = (180–30)/2 = 75. Thus ∠ABQ = 70–75 = 15. Hence point Q is the same as point P, and the assertion is proved.

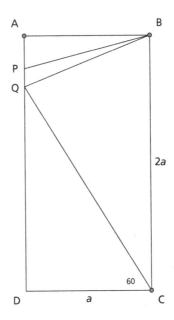

Figure 12.11

What Vesna proposes here meets my objection to Tony Pay's "final" solution in our original correspondence, which I paste here again for ease of reference:

> [Tim to Tony P.] This is very neat, but I'm still stuck! I see that if CP = CD and CD = 2AD then it follows that ADP = 15–the crucial insight being your angle PDA = 1/2 BPC. But (and I hope this isn't being pedantic) I can't deduce CP = CD from the premises: ADP = 15 and CD = 2AD.

In many respects the two solutions—Tony P's and Vesna's—are very similar. Vesna defines Q by constructing an angle of 60 at C, whereas Tony, in effect, defines P as the intersection of AD and the circle through B with centre C (although the points were labelled differently then). In both cases, CDQ is a half an equilateral triangle, and CQB isosceles with base angles of 75. Vesna's logical trick is the "identical points" strategy, a possibility that I had not recognised in my message to Tony P. (the one pasted immediately above).

TWO MORE PROOFS

This is the moment to explain the very nice proof by one teacher mentioned earlier, reported in (Zaslavsky, 2005). The version given here is the general case.

Referring to Figure 12.1: let $\alpha = 60 + \delta$, where $\delta \neq 0$. Then $\angle COD = 60 - 2\delta$, which is less than α if $\delta > 0$, and greater than α otherwise. Also, $\angle COB = 75 + \delta$, and $\angle OBC = 75$. Let the side of the square be 1, and OC = x. Suppose that $\delta > 0$. Then by reference to triangle OBC we have $x < 1$. However, by reference to triangle OCD we have $x > 1$. If $\delta < 0$, the two inequalities between x and 1 are reversed, but the contradiction remains.

This is incredibly neat, and undeniably ingenious. The implicit theorem being brought to bear on this situation is (succinctly) "greater angles lie opposite greater sides," and one would want to be assured that this is part of the Euclidean edifice. In fact, Euclid's proposition 1.19 is: if two sides of a triangle are unequal, the angle opposite the longer side is greater than the angle opposite the shorter side. Clement V. Durell (1939) has it as his Theorem 22.

As I neared the completion of this chapter, I realized that I had no record of any proof of my own. Perhaps I didn't have one? In that case, perhaps I should. Returning from a few days' walking on the Isle of Wight, I attempted a proof on a piece of hotel telephone message paper as the train took me from Portsmouth to Waterloo. It struck me that all of the proofs that I had been sent involved a construction of some kind to be added to the original figure.

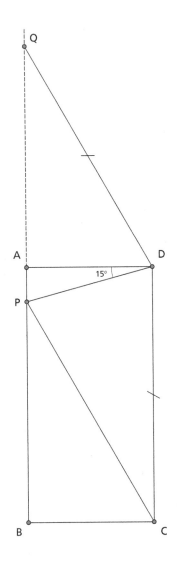

Figure 12.12

My first attempt was to construct a point Q on BA produced such that DQ is parallel to CP. This gives CDQP rotational symmetry (it is a parallelogram) whereas I wanted line symmetry about DP. My proof now proceeds as follows:

Construct a point Q on BA produced such that DQ is equal to CP (Q is the intersection of BA produced with the circle centre D, radius DC). Now triangle QAD is right-angled with hypotenuse twice one adjacent side, so ∠ADQ = 60. [As remarked by others earlier, we could argue that QAD is half of an equilateral triangle]. Therefore $\angle PDQ = \angle PDC = 75$, and so triangles PDC, PDQ are congruent (SAS). Hence $\angle DPC = \angle DPQ = (90 - 15) = 75$. Finally $\angle BPC = 180 - (75 + 75) = 30$ and $\angle BCP = 60$.

I observe my own reluctance to use the identical points strategy. My proof is very direct. I lack the nerve, or the vision, of my correspondents.

Before I left for the walking trip, I had emailed my correspondents a draft version of this chapter, seeking their approval and inviting comments. When I returned, there was a message from Tony Barnard commenting on Del Smith's circle proof ("it's never too late"):

> I looked at Del's proof on the tube into College today. Now, the key to the problem seems basically to be an equilateral triangle, or half of one. Bringing in circle theorems (something I usually like) is a bit of a smokescreen in this case, as essentially all that is used is the isosceles triangles aspect (of the standard proof of 'angle at centre = twice angle at circumference'). So Del's proof can accordingly be lightened and such a version is attached. If one gives weightings to steps in a proof according to how many steps down they are on our flow chart (talking very vaguely here!), this one and the one I sent on 12th April[8] are probably the two of least weight (so far!).

The proof that Tony had attached was essentially the one that I had constructed on the train from Portsmouth, the only difference being that his proof began by identifying Q as the point (on BA produced) equidistant from P and D. I was beginning to feel a sense of closure.

I returned later to John's (Tony B's friend) solution no. 2 (Figure 12.5), because an incorrectly-labelled diagram (all my own work...) had caused me to struggle with it. The proof is essentially based on one circle with centre B, and another ("by symmetry") with centre C. I drew them. Eureka! Join the centres B, C to each other, and to one of the points of intersection (E) of the two circles (Figure 12.13).

Triangle BCE is equilateral, each side being a radius of at least one of the circles. Triangles BEA and DED are isosceles, for the same reason. Simple angle-chasing (or the alternate segment theorem) then establishes that $\angle EAD$ and $\angle EDA$ are both 15. Now I'm sure that that is how the problem came to be set in the first place—doodling with the intersecting circles, like that hexagonal "flower" pattern we drew as kids, playing with compasses. Now strip away the circles and we are left with Figure 12.1! Before, everything was obvious, but now the relationships between the elements of the diagram are veiled. This is not to say that Figure 12.13 (and my commentary on it) *is* a solution of The (original) Problem, but it could be used to construct an 'identical points' proof: in Figure 12.13, let E′ be the point such that $\angle E'AD = \angle E'DA = 15$. Then it will follow that

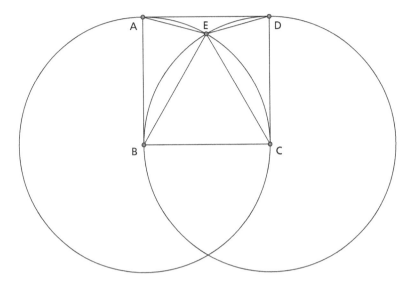

Figure 12.13

∠E′AE = ∠E′DE = 0, and so E′ lies on both DE and AE (produced if necessary), and E is the unique point with that property.

CONCLUSION: CREATIVITY, ELEGANCE, AND WEIGHING PROOFS

These multiple solutions of The Problem illustrate well the scope for creativity offered by Euclidean geometry within the confines of relatively elementary mathematics, and it is a pity that much of this has been lost from the curriculum in England. The same is true, incidentally, of conics in coordinate geometry, where problems about intersections of tangents and normals to ellipses with their axes offer possibilities of working with equations or parametric forms, for example. Creativity is an elusive and over-hyped notion in school mathematics (Huckstep & Rowland, 2001), but it can legitimately be associated with flexible and original approaches to problem solving. If these abilities can be learned, or at least acquired, geometry seems to be a good training ground in which to develop them.

Various aesthetic judgements and expressions of affect are evident in the correspondence about The Problem, from Tony Pay's search for a "nice" solution to Vesna's transparent expression of her feelings about the problem itself. I like Tony Barnard's notion of 'weighing' proofs according to how far they plunder the interconnected structure of the theorems of Euclidean geometry (or to reverse the "steps down" metaphor, how high they reach into the logical edifice). Tony's reference to "our flow chart" is to what we had called The Map in our curriculum development project—a network of Euclidean theorems, displayed and partially ordered by logical dependence: a deductive family tree, as it were, originating in the axioms. Part of The Map is shown as a figure in Barnard (2002). By this criterion, the teacher's proof by contradiction, drawing on Euclid's proposition 1.19, would be relatively heavy: the theorem is on The Map, and in the bottom half.

This preference for "light" proofs indicates, it seems to me, appreciation of the mathematician who brings a minimal toolkit to his work—who likes to travel light, in fact. Parallels come to mind in music and in fine art. I wonder whether, in mathematics, this appreciation of the light touch is in conflict with the elusive notion of "elegance"—the pleasing and seemingly-economical production of the argument. I can best illustrate the same notion by reference to a topic in Number Theory.[9] In 1640, Fermat proposed that every prime number of the form $4k + 1$ can be expressed as a sum of two integer squares. Euler's renowned 1747 proof "by descent" could be said to be "light". Some determination is needed to follow it through, but no results are used that cannot be readily demonstrated. By contrast, the 1770 proof of Laplace brings the whole machinery of the theory of quadratic forms to bear on the problem. It argues that every prime of the given form is represented by *some* quadratic form with "discriminant" −4, that $x^2 + y^2$ has discriminant −4, and that the discriminant −4 has "class number" 1. Fermat's proposition emerges apparently painlessly, as a mere corollary of the class number result. Is that elegant, or is it too "heavy"? I have discussed this question with my students, although their attention is more likely to be focused on which proof to learn for the exam.

I have enjoyed sharing the fun of doing the mathematics along with my email correspondents. My story illustrates once again the place of mathematics for mathematics' sake in the private lives of individual human beings, expressed in their willingness—eagerness even—to share their pleasing insights (or are they revelations?) with others. It shows, once again, that there are reasons for teaching and learning mathematics that lie beyond utility (Ernest, 2000), and perhaps beyond rational argument. I found the following quotation from Poincaré in the Preamble to David Wells'[10] recent book (Wells, 2008, p. 9)

> Those skilled in mathematics find in it pleasures akin to those which painting and music give. They admire the delicate harmony of numbers and of forms; they marvel when a new discovery opens an unexpected perspective; and is this pleasure not esthetic, even though the senses have no part in it?

This isn't quite the end of my story, and indeed I hope that there will be more to tell at another time. It would be good to tell you more, Paul, but, as Andrew Wiles famously said, I think I'll finish there.

ACKNOWLEDGEMENTS

I acknowledge with thanks the inspiration of Orit Zaslavsky, and the contributions of Tony Pay, Tony Barnard, Rex Watson, Derek Smith, Vesna Kadelburg and David Wells: without which, as should be very apparent by now, this chapter could not have been written.

NOTES

1. I will return to one of these proofs later in the Chapter.
2. The email transcripts are nearly always verbatim; the few deviations from this rule are in the cause of brevity and clarity.
3. An online "source for used, new, rare and out-of-print books".
4. Those students who achieve a First Class pass in Part II of the Mathematical Tripos at the University of Cambridge are called 'wranglers'. The ranking of wranglers ceased in 1909.
5. Examinations taken until relatively recently by candidates for the award of Scholarships, Exhibitions and ordinary entrance to the Colleges of the University of Cambridge.
6. I had a further insight into this solution much later: I have described it towards the end of this chapter—in connection with Figure 12.13.
7. As a Sussex alumnus yourself, you may well have come across him, Paul.
8. The proof that Tony had sent on 12th April (in reply to a message from me asking permission to quote from our original correspondence) was essentially the same as 'Bob's proof' (Figure 12.2) and Tony Pay's most recent one.
9. The topic is one from a course that I inherited from Paul, in fact (Burn, 1982).
10. This came to my attention as a consequence of very recent correspondence with David about The Problem!

REFERENCES

Barnard, A. (2002) Starting points and end points. *Mathematics in School 31*(3), 23–26.

Burn, R. P. (1982) *A pathway into number theory*. Cambridge University Press.

Durell, C. V. (1920). *Modern geometry: The straight line and the circle*. London: Macmillan.

Durell, C. V. (1939) *A new geometry for schools*. London : G. Bell & Sons.

Ernest, P. (2000). Why teach mathematics? In S. Bramall & J. White (Eds.), *Why learn maths?* (pp. 1–14). London: Institute of Education.

Huckstep, P., & Rowland, T. (2001) Being creative with the truth? Self-expression and originality in pupils' mathematics'. *Research in Mathematics Education 2*, 183–196.

Linderholme, C. E. (1971). *Mathematics made difficult.* London: Wolfe.

Roth, L. (1948). *Modern elementary geometry.* London: Nelson.

Rowland, T. (2003) Mathematics as human activity: a different handshakes problem. *The Mathematics Educator 7*(2), 55–70.

Wells, D. (2008) *What's the point?* London, Bristol: Rain Press.

Zaslavsky, O. (2005). Seizing the opportunity to create uncertainty in learning mathematics. *Educational Studies in Mathematics 60*, 297–321.

CHAPTER 13

NEEDS VERSUS DEMANDS

Some Ideas on What It Means to Know Mathematics in Society

Tine Wedege
Malmö University, Sweden

Why do we have to learn the division of two fractions? This is a simple question from students, nevertheless causing trouble to mathematics teachers all over the world. But, do they really have to be prepared for answering specific questions like this about any mathematical concept, formula and method? As an educator, I find it important that future mathematics teachers know about the justification problem and are prepared to discuss the global problem: Why teach—and learn—mathematic at all. I see this as a prerequisite for going into a dialogue with pupils about local problems like: Why learn to divide fractions in lower secondary school in Sweden? Any answer to the global problem providing reasons for teaching mathematics may be economic, cultural, technological, political, ideological, historical etc. (Jensen, Niss, & Wedege, 1998). Thus, any answer is value based and, furthermore, any reason given is based on an idea of the content in mathematics education. It is not possible to reflect upon the question why

Relatively and Philosophically Eᵘrnest, pages 215–228

without engaging with the question what and vice versa. "Why" is tangled with "what" in mathematics education and, in any discussion about reasons for teaching and learning mathematics, the issue of content is present, explicitly or implicitly. Thus, the philosophical issue behind any justification debate in mathematics education is about the nature of mathematics and on knowing mathematics. That is the reason why the teacher students in Malmö read Paul Ernest's article, *Relevance versus Utility: Some ideas on what it means to know mathematics* (Ernest, 2004).

In this chapter, after an introduction to the justification problem, I will illustrate the complexity of this problem in mathematics education by presenting and discussing the dualism between utility and relevance, as Ernest is seeing it in a social context, and subsequently the dualism between demands and needs from a sociomathematical point of view. From the discussion of what does it mean to know mathematics, I move to some ideas of what it means to know mathematics in society.

JUSTIFICATION IN MATHEMATICS EDUCATION

The term "justification problem" (Danish: begrundelsesproblemet) was introduced by Skovsmose in 1980 to describe difficulties faced by a group of teacher students when they had to legitimate mathematics at a parents' meeting, according to Johansen (2006). In an international context, Niss (1996) has studied the justification and provided an analysis of its nature in the handbook chapter "Goals of mathematics teaching". Previously, he had pointed to the need of a terminological distinction between two kinds of goals: The term "purposes" includes the motives and reasons which justify the existence of mathematics education. While the term "objectives" includes the specific qualifications (today: competencies) which mathematics education is aiming at (Niss, 1981). The *justification problem* deals with the purposes, the reasons, motives and arguments, for providing mathematics education to a given category of students. Answers to the question "why mathematics education?" are in focus and they have to "rely on and reflect perceptions of the role of mathematics in society, of the philosophy of mathematics, the socioeconomic and cultural structure, conditions and environment in society, ideological and political ideals, and thus vary with place and time" (Niss, 1994, p. 373).

The purposes—motives and reasons—for mathematics education are seldom on the agenda in the public debate. When politicians and educators argue explicitly for the need of teaching and studying mathematics in society it can be seen a sign of a crisis in the school subject. From the end of the 1980s, a trend of decreasing enrolment in education involving mathematics and physics was observed, internationally. This was the background for a conference on justification problems held in Denmark 1997 (Jensen, Niss, & Wedege, 1998). As one of the speakers at this conference, Ernest described the justification problem like this:

> The justification problem in mathematics education concerns the following questions. Why teach mathematics? What is the philosophy of mathematics education in terms of the purposes, goals, justifications, and reasons for teaching mathematics? How can current plans and practices be justified? What might be rationale for reformed, future or possible approaches for mathematics teaching? What should be the reason for teaching mathematics, if it is taught at all? These questions begin to indicate the scope of the justification problem.... (Ernest, 1998, p. 33)

It is obvious from both definitions that answers to the question "Why mathematics education" rely on and reflect conceptions of the role of mathematics in society and positions in philosophy of mathematics education. In his further analysis, Niss (1996) takes an educational, societal and political glance at the problem by grouping the (official) reasons for mathematics education in three general types: (a) Contributing to the technological and socio-economic development of society at large; (b) Contributing to society's political, ideological and cultural maintenance and development; (c) "Providing individuals with prerequisites which may help them to cope with life in the various spheres in which they live: education or occupation; private life; social life; life as a citizen" (Niss, 1996, p. 13).

In a debate on reasons and motives for mathematics education, there are two main types of purposes at two agent levels: one referring to the needs of society (general) and one to the needs of individuals (subjective) and doing this explicitly or implicitly. Furthermore, the focus can be global concerning mathematics education as such or local dealing with the question why mathematics education for this category of students in this specific educational programme. Without setting this up explicitly the dialogue is complicated. In the conference book mentioned above, a justification matrix with two dimensions is offered for reflections and analyses of reasons for teaching and studying mathematics (see Table 13.1).

In the matrix in Table 13.1, four different points of view are possible. By *general* reasons we mean reasons that reside on either a societal or an institutional level, determined from "outside" or by the "system." From a general viewpoint, the focus can be on the overall reasons for the very presence of mathematics in a variety of educational programmes (1). Such global reasons may be economic, cultural, technological, political, ideological, historical etc. The focus can also be on the local reasons for the design, organization, and implementation of programmes in specific educational sectors or institutions (3). By *subjective* reasons we mean reasons that reside on the level of the individual. From a subjective viewpoint, the issues are

**TABLE 13.1   Justification Matrix
(after Jensen, Niss & Wedege, 1998, p. 10)**

Extent \ Agent level	General ("system") reasons[a]	Subjective ("individual") reasons
Global reasons	General reasons for the very existence of studies involving mathematics (1)	Subjective reasons for engaging in studies involving mathematics at all (2)
Local reasons	General reasons for the specific design, organisation, and implementation of specific programmes (3)	Subjective reasons for engaging in particular aspects and activities of a programme in particular ways (4)

[a] In (Jensen, Niss & Wedege, 1998) the term used in this column is "objective." We stated that the label "objective" did not imply any consideration of the validity of these reasons. The distinction between objective and subjective reasons for teaching and learning—and the two terms "objective" and "subjective"—can be seen in the context of the research and debate on vocational qualifications in the 1980s and the the 1990s where the "subjective" dimension of qualifications in terms of personal traits/attitudes such as precision, solidarity, flexibility and the ability to cooperate came into focus. In my research on adults' mathematics containing competences in work, I have claimed that two different lines of approach are possible and necessary: the *objective approach* (the labour market's requirements with regard to mathematical knowledge), and the *subjective approach* (adults' need for mathematical knowledge in their present and future workplace) (see Wedege, 2000). However the term "objective" is very often understood as "neutral" or "un-biased," in debates with researchers. Thus, I have decided to change the term "objective" into "general" in Table 13.1 and later in the discussion of needs versus demands as well.

about the (global) reasons that make a given individual decide—at all—to involve in mathematical studies (2), and about the (local) reasons for an individual to engage in specific aspects of these studies (4) (Jensen, Niss & Wedege, 1998).

When analysing the goals (purposes and objectives) of mathematics education from the general point of view as a reflection of the needs of a given society there are some methodological difficulties. The researcher faces the principal and practical problem of detecting and identifying (and documenting) the actual impact of mathematics education of social needs (see Niss, 1981). Nevertheless, scientists like Kilpatrick, Swafford, & Findell (2001, p. xiii) and Bishop (1999, p. 1) argue for societies' "mounting need for proficiency in mathematics" and demands of "much greater mathematical knowledge" respectively and doing this without any references to empirical evidence. Skovsmose (2006) finds that much research in mathematics education is based on the assumption that mathematics education contains an intrinsic value. He refers to an essentialism assuming that there is a positive value in mathematics education guaranteed by the very fact that this education addresses mathematics. Skovsmose points to this assumption as ensuring that "mathematics educators can operate as "ambassadors" of mathematics, with the certainty that they are acting on behalf of a good cause" (p. 276). In his justification article mentioned above, Ernest (1998) discusses myths about mathematics in society; for example that "mathematics is very useful, and more maths skills are needed among the general populace in industrialized societies". He states, what he sees as his most controversial set of claims, "The utility of mathematics in the modern world is greatly overestimated, and the utilitarian argument provides a poor justification for the universal teaching of the subject throughout the years of compulsory schooling." (Ernest, 1998, p. 38).

As asserted above, it is appropriate to make a terminological distinction in mathematics education goals between the *purposes* (motives, reasons for teaching and learning mathematics) and the *objectives* (mathematical capability, proficiency, knowledge, competence, skills, etc. as aims in the teaching and learning of mathematics). From this follows that one can see why (reason) and what (content) in mathematics education as two sides of the same coin.

OBJECTIVES: RELEVANCE VERSUS UTILITY

Stated in general terms, the objective of mathematics education is that students know mathematics. But, what does it mean—or might it mean—to know mathematics? This is the central issue in Ernest (2004) where he uses the term "mathematical capability"—and not "mathematical knowledge—to present and discuss his ideas in the social context of mathematics teaching and its individual outcomes. In this discussion, Ernest puts aside differences between societies and cultures and he uses a post-industrial western democracy like Sweden as his reference point. The students in his reflections are 16–18 years of age and about the end of compulsory schooling. He is doing this with a focus on general and common issues without looking at different individual needs. As any discussion about objectives this one is inevitably related to the purposes of mathematics education and—with the generalisations mentioned—Ernest puts himself at the global level combined with the subjective agent level: individual reasons for engaging in studies involving mathematics at all (2 in the justification matrix, Table 13.1).

Ernest claims that human capabilities and capacities are always for certain activities, purposes and functions. But, in order no to fall into a utilitarian trap, he draws a distinction between "utility" and "relevance". He finds that both qualities are value-laden terms used to denote what is the speaker deems apposite relative to a given context:

Utility means "a narrowly conceived usefulness that can be demonstrated immediately or in the short term, without consideration of broader contexts or longer term goals" (p. 314). One of the outcomes of utilitarian education is that mathematics is communicated as narrowly technical isolated from issues of value and social concern. *Relevance* is seen as a ternary relation between three things (R, P, G). R is a situation, an activity or an object to which relevance is ascribed. P is a person or group of people who ascribes relevance to R. G is a goal which embodies the values of P in this instance. Thus object R is relevant when considered so by the person P in achieving the goal G. For example school mathematics (the object R) is said to be relevant "by many politicians and educational leaders (the group P) with the aim (the goal G) of increasing the mathematical competence and technologically-related employment skills of the population (which is assumed to increase economic output and national prosperity) (p. 315). In his clarification of what relevance means in mathematics education, the term used by Ernest is "goals". Even though this is not explicit, it seems that the term, in the context of the relevance discussion, encompasses both meanings (purpose and objective).

In order to answer the opening question (what it means to know mathematics), Ernest distinguishes a series of different objectives for teaching and learning mathematics based on an analysis that he has made of five ideological groupings in the mathematics curriculum debate in Britain (industrial trainer, technological pragmatist, old humanist, progressive educator, public educator) (Ernest, 1991). Corresponding to these objectives, which Ernest sees as complementary, he formulates five capabilities to be developed by the learners and added a sixth capacity: the appreciation of mathematics in itself as an element of culture (see Table 13.2).

Internationally, the logic of competence has taken over in the educational discourse in general and in the discourse in mathematics education specifically, since the mid 1990s (Wedege, 2003b). Ernest does not use the term "competence", but in his construction of mathematical capabilities and capacities, the main ideas of competence as objective in educations are incorporated.

According to Ernest (2004), the learning objectives in Table 13.2 are not mutually exclusive and they are not necessarily desirable or relevant for all learners. I see the six objectives formulated as capabilities or capacities forming together a "mathematical competence" (a construction of what it might mean to be mathematical competent) as an answer to the question what does it mean to know mathematics. The objectives "utilitarian knowledge" (1) and "practical, work-related knowledge" (2) are both presented as the ability to do something (knowing how), and they are supposed to be useful for the students as future workers. "Advanced specialist knowledge" (3) is supposed to be useful for some and relevant to others (3) and "appre-

TABLE 13.2 Different Teaching and Learning Objectives and Capabilities

Objective	Associated mathematical capabilities
1. Utilitarian knowledge	To be able to demonstrate useful mathematical and numeracy skills adequate for successful general employment and functioning in society
2. Practical, work-related knowledge	To be able to solve practical problems with mathematics, especially industry and work centered problems
3. Advanced specialist knowledge	To have an understanding and capabilities in advanced mathematics, with specialist knowledge beyond standard school mathematics (...)
4. Appreciation of mathematics	To have an appreciation of mathematics as a discipline including its structure, subspecialisms, the history of mathematics and the role of mathematics in culture and society in general
5. Mathematical confidence	To be confident in one's personal knowledge of mathematics, to be able to see mathematical connections and solve mathematical problems, and to be able to acquire new knowledge and skills when needed
6. Social empowerment through mathematics	To be empowered through knowledge of mathematics as a highly numerate critical citizen in society, able to use this knowledge in social and political realms of activity

Source: Ernest, 2004 p. 317

ciation of mathematics" (4) is supposed to be relevant to all students (knowing that). The two last objectives "mathematical confidence" (5) and "social empowerment through mathematics" (6) represent affective and social dimensions of being mathematical competent. Thus, Ernest's construction of mathematical competence in a social context encompasses cognitive elements (skills and knowledge), affective elements (confidence) and social elements (empowerment) as well as their dynamic interplay.

TO KNOW MATHEMATICS IN SOCIETY

The context for Ernest's construction of mathematical knowledge—as capabilities and capacities—is social. Concepts of people's social competences like ethnomathematics and folk mathematics, as well as concepts of adult numeracy, mathematical literacy and of mathemacy, have expanded the problem field of mathematics education research (Gerdes, 1996; Jablonka, 2003; Mellin-Olsen, 1987; Skovsmose, 1994; Wedege, 1999). Today it is scientifically legitimate to ask questions concerning people's everyday mathematics and about the power relations involved in mathematics education. In other words, it is legitimate to ask "What does it mean to know mathematics in society?" In the studies of for example ethnomathematics and adult numeracy, a social approach is common and a critical perspective can be opened up when studies concern the functions of mathematics education in society and in people's lives.

By definition, any notion of mathematical literacy or numeracy contains the societal dimensions of mathematics, technology and culture. The definitions of mathematical literacy and of numeracy and the related studies are concerned with the relationships between people, mathematics and society. On the basis of previous studies, I have given a preliminary definition of an analytical concept, which encompasses the studies of numeracy and mathematical literacy and related concepts of knowing mathematics in society in a single term. By *sociomathematics*,[1] I mean:

- A problem field concerning the relationships between people, mathematics and society, and
- A subject field combining mathematics, people and society—as we may find it for example in ethnomathematics, folk mathematics or adult numeracy (Wedege, 2003a, p. 2).

In my terminology, sociomathematics is the name of a subject field (a field of study) and a specific problem field just like ethnomathematics (see Gerdes, 1996) (see Figure 13.1).

Sociomathematical problems concern (1) people's relationships with mathematics (education) *in* society and vice versa. The relationship between hu-

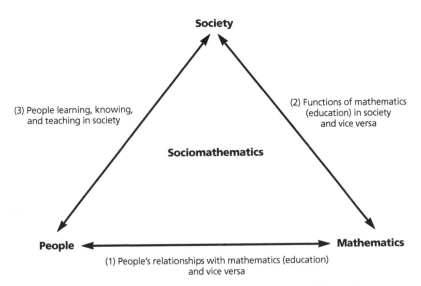

Figure 13.1 Sociomathematics as a subject field (Wedege, 2003a, p. 2).

mans and mathematics can be seen as cognitive, affective or social according to the given view point of a specific study. This relationship is the key issue, but to investigate this problem one has to study two other problems: (2) the functions of mathematics (education) in society and vice versa, and (3) people learning, knowing and teaching *in* society. Power is a central sociomathematical issue related to all three dimensions. In his book "The Politics of Mathematics Education", Mellin-Olsen (1987) stated that it is a political question whether folkmathematics is recognized as mathematics or not. He presents the book as a result of a twenty year long search "to find out why so many intelligent pupils do not learn mathematics whereas, at the same time, it is easy to discover mathematics in their out-of-school activities" (p. xiii). FitzSimons (2002) states that the distribution of knowledge in society defines the distribution of power and, in this context, people's everyday competences do not count as mathematics. In policy documents in educational systems, in teachers' practices, and in research in the teaching and learning of mathematics, the power of mathematics and mathematics education is clearly assumed (Skovsmose & Valero, 2002). However, it is not clear what is really meant by the terms "power" and "mathematics," particularly when it is being used differently by the multiple actors involved in giving meaning to the practices of the teaching and learning of mathematics in society (Valero & Wedege, 2009).

In Skovmose's studies of students' learning obstacles in mathematics, one finds an example of a *sociomathematical concept* construction. He does not find the cultural background of the students sufficient to account for the situation but also involves their *foreground*, i.e., the opportunities provided by the social, political and cultural situation: "When a society has stolen away the future of some group of children, then it has also stolen the incitements of learning" (Skovsmose, 2005, p. 6). An example of a *sociomathematical study* is found in my inquiry of adults learning mathematics (Wedege, 1999). I go beyond the local situation given by Lave's socio-psychological concept of community of practice and involve Bourdieu's sociological concept about *habitus* meaning a system of dispositions which allow the individual to act, think and orient him or herself in the social world. People's habitus is incorporated in the life they have lived up to the present and consists of systems of durable, transposable dispositions as principles of generating and structuring practices and representations (Bourdieu, 1980).

PURPOSE: NEEDS VERSUS DEMANDS

In a study of knowing mathematics in society (mathematical literacy/numeracy) two different lines of approach are possible and intertwined in the research: a *subjective approach* starting with people's subjective needs in their societal lives, and a *general approach* starting either with societal and labour market demands and/or with the academic discipline mathematics (transformed into "school mathematics"). The general approach is obviously to be found in international surveys on mathematical literacy and numeracy like OECD (2005, 2006). This approach is also represented by the famous "Cockcroft report" with a large-scale British investigation of the mathematical needs of adult life initiated in order to make recommendations concerning the curriculum in primary and secondary schools. The title of this report was "Mathematics counts" (Cockcroft, 1982). Fifteen years later "Adults count too" was the title of a book written by Benn (1997). As the titles of the two books suggest the approaches are different. Benn started with the adults. She argued that mathematics is not a value-free construct, but is imbued with elitist notions which exclude and mystify. She recognises but rejects the discourse of mathematics for purposes of social control, where mathematical literacy is seen as a way of maintaining the status quo and producing conformist and economically productive citizens (Benn, 1997). However, to understand the affective and social conditions for people's learning processes in mathematics one has to take both dimensions into account (see Wedege, 2000; Wedege & Evans, 2006). And five years later FitzSimons (2002) published the book "What counts as mathematics?" In her discussion of technologies of power in adult and vocational education one may find the dialectic between the two approaches—the general and the subjective.

In this section, my starting point is two concepts of knowing mathematics in society: mathematical literacy and numeracy.

In policy reports, international surveys, in national curricula and in research, the term "mathematical literacy" pretends to provide an answer to the question about knowing mathematics in society and does this in the terms of competencies. As with the ideas of literacy, there is a common notion of functional knowledge and of situatedness across the different constructions of concepts of mathematical literacy and of numeracy, which is crucial to the logic of competence. However, the different concepts of knowing mathematics in society (mathematical literacy) are based—implicitly or explicitly—on different notions of human knowledge and of learning mathematics (Wedege, 2003b), and they vary with the approaches, values and rationales of the stakeholders and researchers (Jablonka, 2003).

In the ALL (the Adult Literacy and Life Skills Survey), one finds a very short and still very informative definition of numeracy: "*Numeracy*—the knowledge and skills required to effectively manage the mathematical demands of diverse situations" (OECD, 2005, p. 16). The approach is general and this is about demands and requirements from society and mathematics. It is obvious from the ALL report that the ideology behind the survey is about people's ability to adapt to changes in technology and society. In

PISA (Programme for International Student Assessment), the general definition reads like this:

> *Mathematical literacy* is an individual's capacity to identify and understand the role that mathematics plays in the world, to make well-founded judgments and to use and engage with mathematics in ways that meet the needs of that individual's life as a constructive, concerned and reflective citizen. (OECD, 2006, p. 72)

According to this definition, the approach of PISA, which pretends to assess mathematical literacy of students near the end of compulsory education, should be subjective and starting with the needs of the individuals. However, the concrete construction of the eight mathematical competencies composing mathematical literacy (thinking and reasoning; argumentation; communication; etc.) is general starting with mathematics and ending up with mathematics. In the test items the so-called real world situations are only a means for re-contextualising mathematical concepts and in the end "it is not the situations themselves which are of interest, but only their mathematical descriptions" (Jablonka, 2003, p. 81).

In a broad definition of numeracy, we have stressed the importance of the societal context (Lindenskov & Wedege, 2001). *Numeracy* is described as an everyday competence—in terms of functional mathematical skills and understanding—that all people in principle need to have in any given society at any given time (p. 5). In our definition, numeracy is thus historically and culturally determined and it changes along with social change and technological development: numeracy in Denmark 2009 might be different from numeracy in Togo 1979. On the other hand, the definition can be seen as an attempt to combine the general and the subjective dimension. The term "in principle" makes possible a general evaluation of numeracy in the population (as in the big international surveys) and the developing of general courses in numeracy. However, all individuals who participate in a numeracy course will, in fact, have their own subjective perspectives (why am I here), their own backgrounds and needs (what am I going to learn) and their own strategies (what am I learning).

Within the construction of goals in mathematics education, there is a dualism between purposes and objectives. It is actually impossible to discuss one side of the coin without taking the other into account. In the section above, the focus was the objectives (mathematical capabilities and capacities as aims in teaching and learning mathematics). However, the pair relevance/utility ensured the link to the purposes (motives, reasons for teaching and learning mathematics). In this section, the focus is the purposes with the pair need/demands creating a link to the objectives.

CONCLUSION

The student asked "Why do we have to learn the division of two fractions?" But maybe he or she was really meaning "Why do I have to learn this?" What seems to be a general justification question (what is the reason for this specific part of the mathematics curriculum) is perhaps a subjective justification question (why should I engage in studying this particular aspect of mathematics). To open up to the students' individual perspectives—or foregrounds—could be the teacher's first step in a debate of goals (objectives and reasons) in the mathematics classroom. Ernest (2004) claims that there is an important perception of relevance missing from the discussion of goals in mathematics: the learners' own views of school mathematics and its relevance to their personal goals. He points out that students' beliefs of mathematics often reflects the dominant rhetoric about the importance and high valuation of mathematics in society—not personal relevance to their own lives. According to Niss (1994), there is a contradiction between the general relevance of mathematics in society and the subjectively experienced irrelevance. He calls this "the relevance paradox" and locates the cause in the "discrepancy between the objective social significance of mathematics and its subjective invisibility" (p. 371). In the light of the discussion of relevance versus utility and of needs versus demands in this chapter, one might ask if this conflict is a paradox or just a matter of a simple confusion of relevance and utility.

REFERENCES

Benn, R. (1997). *Adults count too: Mathematics for empowerment.* Leicester: NIACE.

Bishop, A. J. (1999). Mathematics teaching and values education: an intersection in need of research. *ZDM, 99*(1), 1–4.

Bourdieu, P. (1980) *Le sens pratique.* Paris: Les éditions de minuit.

Cobb, P. (1996). Accounting for mathematical learning in the social context of the classroom. In C. Alsina et al. (Eds.), *8th International Congress on mathematical education: Selected lectures,* (pp.85–99). Sevilla: S.A.E.M. 'THALES'. (Published in 1998.)

Cockroft, W. H. (Chairman of the Committee of Inquiry into the Teaching of Mathematics in Schools) (1982). *Mathematics counts.* London: Her Majesty's Stationery Office.

Ernest, P. (1991). *The philosophie of mathematics education.* London: Falmer Press.

Ernest, P. (1998). Why teach mathematics? The justification problem in mathematics education. In J. H. Jensen, M. Niss, & T. Wedege (Eds.), *Justification and enrolment problems in education involving mathematics or physics,* (pp. 33–55). Frederiksberg, Denmark: Roskilde University Press.

Ernest, P. (2004). Relevance versus utility: some ideas on what it means to know mathematics. In B. Clarke et al. (Eds.), *International perspectives on learning and teaching mathematics,* (pp. 313–327). Göteborg, Sweden: National Center for Mathematics Education, NCM.

FitzSimons, G. E. (2002). *What counts as mathematics? Technologies of Power in Adult and Vocational Education.* Dordrecht, The Netherlands: Kluwer Academic.

Gerdes, P. (1996). Ethnomathematics and mathematics education. In A. J. Bishop et al. (Eds.), *International handbook of mathematics education* (pp. 909–943). Dordrecht, The Netherlands: Kluwer Academic.

Jablonka, E. (2003). Mathematical literacy. In A. J. Bishop et al. (Eds.), *Second international handbook of mathematics education* (pp. 75–102). Dordrecht, The Netherlands: Kluwer Academic.

Jensen, J. H., Niss, M., & Wedege, T. (Eds.) (1998). *Justification and enrolment problems in education involving mathematics or physics.* Frederiksberg, Denmark: Roskilde University Press.

Johansen, L. Ø. 2006. *Hvorfor skal voksne tilbydes undervisning i matematik? En diskursanalytisk tilgang til begrundelsesproblemet.* Doctoral thesis. DCN, Aalborg University.

Kilpatrick, J., Swafford, J., & Findell, B. (Eds.) (2001). *Adding it up: Helping children learn mathematics.* Washington, DC: National Academy Press.

Lindenskov, L., & Wedege, T. (2001). Numeracy as an analytical tool in adult education and research. *Centre for Research in Learning Mathematics, Publication no. 31,* Roskilde University. (Retrievable at http://www.statvoks.no/emma/materials_numeracy.htm)

Mellin-Olsen, S. (1987). *The politics of mathematics education.* Dordrecht: Kluwer Academic.

Niss, M. (1981). Goals as a reflection of the needs of society. In R. Morris (Ed.), *Studies in mathematics education,* (pp. 1–22). Paris: UNESCO.

Niss, M. (1994), Mathematics in society. In R. Bieler, R. W. Scholz, R. Sträßer, & B. Winkelmann (Eds.), *Didactics of mathematics as a scientific discipline,* (pp. 367–378). Dordrecht, The Netherlands: Kluwer Academic.

Niss, M. (1996). Goals of mathematics teaching. In A. J. Bishop et al. (Eds.), *International handbook of mathematics education* (pp. 11–47). Dordrecht: Kluwer Academic.

OECD (2005). *Learning a living: First results of the adult literacy and life skills survey.* Paris: Statistics Canada and OECD.

OECD (2006). *Assessing scientific, reading and mathematical literacy. A framework for PISA 2006.* Paris: OECD.

Skovsmose, O. (1994). *Towards a philosophy of critical mathematics education.* Dordrecht, The Netherlands: Kluwer Academic.

Skovsmose, O., & Valero, P. (2002). Democratic access to powerful mathematical ideas. In L. D. English (Ed.), *Handbook of international research in mathematics education: Directions for the 21st century* (pp. 383–407). Mahwah, NJ: Lawrence Erlbaum.

Skovsmose, O. (2005). Foregrounds and politics of learning obstacles. *For the Learning of Mathematics, 25*(1), 4–10.

Skovsmose, O. (2006). Research, practice, uncertainty and responsibility. *Mathematical Behaviour 25,* 267–284.

Valero, P., & Wedege, T. (2009). Lifelong mathematics education (2): Empower, disempower, counterpower? In C. Winsløw (Ed.), *Nordic research in mathematics education* (pp. 363–366). Rotterdam, The Netherlands: Sense.

Wedege, T. (1999). To know or not to know—mathematics, that is a question of context. *Educational Studies in Mathematics, 39*, 205–227.

Wedege, T. (2000). Technology, competences and mathematics. In D. Coben, J. O'Donoghue, & G. E. FitzSimons (Eds.), *Perspectives on adults learning mathematics: Research and practice* (pp. 191–207). Dordrecht, The Netherlands: Kluwer Academic.

Wedege, T. (2003a). Sociomathematics: people and mathematics in society. *Adults Learning Maths—Newsletter 20*, 1–4.

Wedege, T. (2003b). Konstruktion af kompetence(begreber). *Dansk Pædagogisk Tidsskrift, 03/2003*, 64–75.

Wedege, T., & Evans, J. (2006). Adults' resistance to learn in school versus adults' competences in work: the case of mathematics. *Adults Learning Mathematics: an International Journal, 1*(2), 28–43

Zaslavsky, C. (1973). *Africa counts*. Boston: Prindle, Weber & Schmidt.

NOTE

1. I found my inspiration to the term "sociomathematics" in *sociolinguistics*, i.e. relationships between language and society constituted as a scientific field within linguistics. But there is an important difference: sociomathematics is a field within mathematics education research (studying people's relationship with mathematics in society), not a sub-discipline of mathematics. Previously, the substantive "sociomathematics" has only been used in a meaning very similar to "ethnomathematics." Zaslavsky (1973, p. 7) explains *sociomathematics* of Africa as "the applications of mathematics in the lives of African people, and, conversely, the influence that African institutions had upon their evolution of their mathematics". The adjective "sociomathematical" is used at the level of the social context of the classroom where Cobb and his colleagues developed the *sociomathematical norms* in an interpretive framework for analyzing mathematical activity with a social dimension (classroom social norms, sociomathematical norms and classroom mathematical practices) and a psychological dimension (beliefs about roles and mathematical activity in school, mathematical beliefs and values, and mathematical conceptions) (Cobb, 1996). In this framework the social category of socio-mathematical norms is correlated with the psychological category of mathematics beliefs and values. In my terminology, studies of socio-mathematical norms in a classroom would be called "sociomathematical" only if the students' relationships with mathematics *in* society are explicitly on the agenda. For example related to the students' gender, ethnicity or class.

CHAPTER 14

NEW TECHNOLOGIES IN THE CLASSROOM

Towards a Semiotic Analysis

Ferdinando Arzarello
Università di Torino, Italy

INTRODUCTION

The introduction of new technologies (NT) in the classroom generally produces a very complex situation. This is particularly true for CAS systems, which are very sophisticated tools: see for example the discussion in Artigue (1997). Roughly speaking, a major question when such a new powerful instrument is introduced in a classroom is, "Q0: *Can the software really support a better learning in students (through a suitable didactical design) and how? Which are its strong and feeble features?*"

The paper scrutinizes this naïve question for a specific software, TI-Nspire (TI-Ns; see: www.TI-Ns.com/tools/nspire/index.html) using two different approaches:

- The *instrumental approach,* coming from the so called *cognitive ergonomy* (Verillon & Rabardel, 1995), summarised in §1;

Relatively and Philosophically E^{u}rnest, pages 229–249
Copyright © 2009 by Information Age Publishing

- The *embodiment approach,* coming from cognitive science (see Wilson, 2002), sketched in §2.

Using them it is possible to translate Q0 into two more precise research problems, labelled Q1 and Q2. To face them a suitable model is introduced, the *Space of Action, Production and Communication* (see Arzarello, 2008), which is apt to shape many didactical phenomena produced in the classroom because of the introduction of the NT (§3). The model points out some specific processes of students who are using TI-Ns: they are described with some detail in §§4.1, 4.2, comparing also the correspondent behaviours of students when using another software. The analysis is pursued focusing the different semiotic resources used by the subjects who are learning mathematics using TI-Ns. We get so a first picture of students' learning processes while interacting with NT and this allows to give some precise answers to Q1 and Q2 (§4).

However, to get a complete picture it is necessary to consider also the role of the teacher and not only the interactions between the students and the software. Hence an analysis of her/his so called *semiotic mediation* enters the scene (§5) and is used to suitably frame the results introduced in the first two chapters.

The findings and results here presented are based on the data got through a two-years experimentation[1] made in two Italian secondary schools, where different CAS systems (and particularly TI-Ns) have been systematically used in the class of mathematics.

1. THE INSTRUMENTAL APPROACH

P. Rabardel and P. Verillon introduce the concept of *Instrumental genesis* to describe the process by which an artifact becomes an instrument. It indicates the two directions in which this process takes place: towards the self and towards outside reality. The first meaning of appropriation requires the artifact to be integrated within one's own cognitive structure (e.g., one's existing representations, available action schemes, etc.) that in general, require adaptation. Rabardel and Verillion termed this self-oriented construction "instrumentation." The second meaning indicates that the artifact has to be appropriated to an outside context. Specific ends and functional properties—some not necessarily intended by design—are attributed to it by the user. Adjustments are made to account for goal and operating conditions. Rabardel and Verillon called this "instrumentalization."

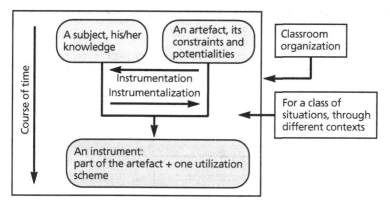

Figure 14.1

Figure14.1 (taken from Maschietto & Trouche, to appear) illustrates this double arrow.

For example, students learn to properly use data-capture in TI-Ns through an instrumentation process. But they generate an instrumentalization process, when they use the potentiality of random number generators in producing samples of data in the spreadsheet for solving a specific fresh problem.

Instrumented action can be broadly distinguished according to whether it aims at producing transformations (*pragmatic action*) or affording knowledge (*epistemic action*).[2] Artifacts (such as sensors, meters, and computers) are also used to derive knowledge concerning the environment by detecting, registering, and measuring some aspect of reality not immediately accessible to the user. Analogously, for the mathematics situations represented in TI-Ns environments (e.g., see Figure 14.2) students may refine their actions for producing a more uniform set of data with data-capture through an animation, for example exploring the relationships among the sides and an angle in a variable triangle, subject to some constraints.

Rephrasing Q0 within the instrumental frame we get the following question:

(Q1): *What is the specificity of instrumental actions in the considered ICT environment? What instrumented actions in this environment help (or block) students learning processes?*

More specifically: what are its instrumentation/instrumentalization aspects, what the interplay between pragmatic and epistemic actions? How the activated instrumented actions do (or do not) support student learning?

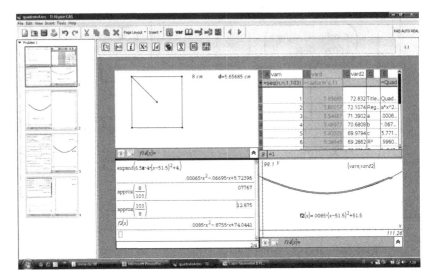

Figure 14.2

2. EMBODIMENT

The second issue concerns *embodiment* and *multimodality*. The notion of *multimodality* has evolved within the paradigm of *embodiment*, which has been developed in these last years (Wilson, 2002). Embodiment is a movement afoot in cognitive science that grants the body a central role in shaping the mind. It concerns different disciplines, e.g., cognitive science and neuroscience, interested with how the body is involved in thinking and learning. The new stance emphasizes sensory and motor functions, as well as their importance for successful interaction with the environment. This is particularly palpable when observing the interactions human-computer. A major consequence is that the boundaries among perception, action and cognition become *porous* (Seitz, 2000). Concepts are so analysed not on the basis of "formal abstract models, totally unrelated to the life of the body, and of the brain regions governing the body's functioning in the world" (Gallese & Lakoff, 2005, p.455), but considering the *multimodality* of our cognitive performances.

Instrumented activity in technological settings is multimodal in an essential way. Many times it reflects the fact that action is not only directed towards objects but is also directed towards persons. Action in such situations can be seen as aiming at altering another subject's state of information. *Communicative action* is of this type, and subjects (teachers and students) simultaneously use a wide array of verbal, gestural, and graphic registers to communicate their thought. All such components or modalities (written signs, oral language, body enactments, artefacts use, etc) intervene in an intertwined way in learning and more in general in knowledge formation. A multimodal mode is typical of people using computer; specifically, many of them allow to use a multifaceted environment, where different representations are simultaneously present and the multimodality of processes is consequently improved and supported. As an example, see the multi-representation features of TI-Ns in Figure 14.2.

Mathematics learning, as it happens in the described context, can be fruitfully analyzed through a semiotic approach that allows us to consider both its cognitive and its didactic dimensions. The resulting semiotic analysis therefore considers a plurality of semiotic resources that goes far beyond the written symbolic systems and oral language, to include gestures and bodily means of expression within the TI-Ns environment.

This frame allows refining Q0 as follows:

Q2: *At what extent does a software modify the usual multimodal behaviours of students?*

The two questions Q1, Q2 can be investigated using a unitary model, the *Space of Action, Production and Communication*, elaborated by Arzarello and his collaborators in these last years (Arzarello, 2008). Through the APC-space all the signs[3] produced or acted on by the different subjects are taken into account. This allows considering both the multimodality of the learning processes that happen in the classroom and the nature of the instrumental actions made by the subjects.

3. THE SPACE OF ACTION, PRODUCTION AND COMMUNICATION

The APC-s describes the processes, which develop and are possibly shared in the classroom among students (and the teacher) while working together. It analyses them considering their different components and a variety of mutually dependent relationships among them.

The components are: the body, the physical world, the cultural and institutional environment in a word, the students themselves and the teacher along with the context where they are acting and learning. When students learn mathematics in an ICT environment, these and other components (e.g., the emotional ones) take an active part in their learning processes, interacting together.

The three letters A, P, C illustrate the main dynamic relationships among such components, namely students' actions and interactions (e.g., in a situation at stake, with their mates, with the teacher, with themselves, with the software), their (individual or collective) productions (e.g., activating a function of the software, answering a question, posing other questions, making a conjecture, introducing a new sign to represent a situation, and so on) and communication aspects (e.g., when the discovered solution is written and communicated orally or in written form to a mate or to the teacher, or sent to the teacher through an electronic platform). The APC-s is a typical complex system, which cannot be described in a linear manner as resulting by the simple superposition of its ingredients. It particularly models how the relationships among its components develop in the classroom through the specific actions of the teacher. The APC-s components analysed through a semiotic lens allow making palpable the main features of the learning processes in technological environments. To grasp them, it is necessary to look at what happens while students are using the software, considering their multimodal productions (not only what they write) in a very detailed way. This is possible only through a careful analysis of the videotaped lectures. In general the students make some (pragmatic or epistemic) actions and produce some signs because of their own personal interpretation they attach to the given task; while the teacher pushes them towards the scientific sense through a careful design of the didactic situation and through suitable use of the artefacts (this strategy is called *semiotic mediation*). Of course there are different interactions between the subjects (students and teacher): hence all of them communicate each other through different semiotic resources. The dynamic interplay among the different components of the APC-space and its intertwining with the in ICT environments are illustrated in Figure 14.3.

4. FIRST ANSWERS TO THE RESEARCH QUESTIONS

I recall here the two research questions discussed above:

Q1: *What is the specificity of instrumental actions in the considered ICT environment? What instrumented actions in this environment help (or block) students learning processes?*

Q2: *At what extent does a software modify the usual multimodal behaviours of students?*

I can now answer them for what concern a specific software, namely TI-Ns. In fact we have carefully analysed teaching and learning processes that happened in two different classes during a two year teaching experiment, where 20 students of grades 9, and 10 and 20 students of grades 12 and 13 attending a scientific oriented school in Italy used systematically

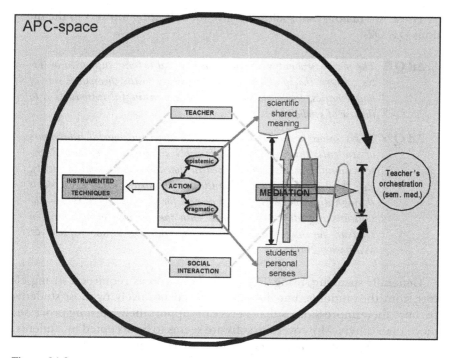

Figure 14.3

TI-Ns software in their classes of mathematics. We videotaped some of their activities (we have more than 100 hours videos), interviewed them, analysed their written protocols produced during their tasks, gave them some tests and questionnaires. All the data were analysed through the APC-space model. We here discuss the two following main findings:

1. A triadic structure seems necessary to describe properly the learning strategies of students who are using TI-Ns: *epistemic, almost-empiric, pragmatic*. A consequence is the productions of fresh practices in mathematics, which positively support learning processes of students. The new structure supports different *tempos* in the way students solve problems, compared with the typical paces with other software.

2. TI-Ns support specific instrumented actions, which push the students towards a meaningful use of symbols, namely new practices are introduced through specific instrumented actions that positively support the *treatment* and the *conversion* of the symbols of mathematics (see Duval, 2006, for this terminology).

These two major results allows to give the following answers to the questions Q1, Q2:

Ad Q1: *The major new entries in the instrumental actions supported by TI-Ns concern the specificity of the transition to the theoretical side of mathematics, to its modeling and to a meaningful introduction to the use of symbols.*

Ad Q2: *TI-Ns seems to modify the tempos of some multimodal behaviours of students; this makes TI-Ns possibly similar to some new Representational Infrastructures[4] used in nowadays technological society, e.g., the increasing habit of simultaneously surfing of youngest people through different technological devices for short period of times—the multitasking attitude- compared with the old way of operating in sequence for longer periods of time (see Baricco, 2006; Tapscott & Willliams, 2007)*

Generally speaking, using TI-Ns in the classrooms requires a strong effort from the students, but the effort is lived positively by most students because their impression[5] is that TI-Ns can support their learning processes more than others. Moreover the software seems to be accepted by students, because of its multitasking aspects, which are consonant with their technological habits outside the school.

4.1 The Triadic Structure of Instrumented Actions within TI-Ns and the Related Tempos

To properly describe TI-Ns instrumented actions I refine the usual dyadic structure epistemic-pragmatic of the instrumental approach and introduce a fresh modality that I have called *almost-empirical* to distinguish it from stricter epistemic ones. To give an idea of what is meant by almost-empirical, I discuss sketchily a typical TI-Ns situation and contrast it with a Cabri-situation.

A simple problem, originated by the PISA test is the following:

Problem 1: *The students A and B attend the same school, which is 3 Km far from A's home and 6 Km far from B's home. What are the possible distances between the two houses?*

A possible solution with TI-Ns is illustrated in Figure 14.4. You draw two circles, whose centre is the school: they represent the possible positions of the two houses with respect to the school. Then you create two points, say a and b, moving on each circle, construct the segment ab and measure it.

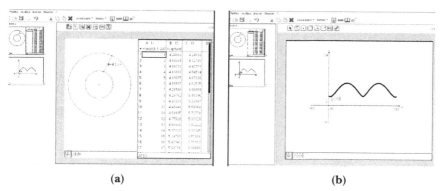

(a)   (b)

Figure 14.4

Successively you create a sequence of the natural number in column A of the spreadsheet (Figure 14.4a) and through two animations (in one you move a and in the other you move b) you collect the corresponding lenghts of ab (columns B, and C in the spreadsheet of Figure 14.4a). In the end you make the scattered plot A Vs/B and A Vs/C (Figure 14.4b). Then you draw your considerations about the possible distances of A's and B's houses, considering the properties of the obtained scattered graph.

Consider now the following problem to solve with Cabri (but it could be given also within TI-Ns):

> **Problem 2:** *Let ABCD be a quadrangle. Consider the perpendicular bisectors of its sides and their intersection points A', B', C', D' of pairwise consecutive bisectors. Drag ABCD, considering its different configurations: what happens to the quadrangle A'B'C'D'? What kind of figure does it become? (Figure 14.5)*

Many students, who are able to use Cabri, after some initial explorations, where they drag almost casually some of the suggested points (*wandering dragging*), almost by case observe that with particular positions, the four points A', B', C', D' coincide. So they try to carefully drag the vertices of the quadrilateral so to keep together the four points A', B', C', D' (*dummy locus dragging*).

After some different explorations they realise that the coincidence situations happens when the four vertices of the quadrilateral have the same distance from the common point A', B', C', D'. Namely when the quadrilateral is cyclic. (For a detailed analysis of this and similar problems see: Arzarello, 2000; Arzarello, Olivero, Paola, & Robutti, 2002).

The two examples are different for many reasons (typology of problem, different difficulties, etc.) but illustrate very well two different approaches:

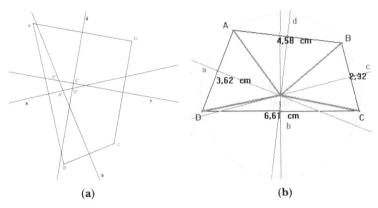

(a)   (b)

Figure 14.5

one is possible both in Cabri and in TI-Ns, the other is typical of TI-Ns but is almost impossible in Cabri for pragmatic reasons.

Let us consider the role of the variable points and the ways they are manipulated in the two cases. For both the Instrumental approach would speak of epistemic instrumented actions. But considering carefully the nature of the instrumented actions in the two environments one finds interesting differences.

In the example of TI-Ns (which is an emblematic example within this environment), TI-Ns allows a collection of data very similar to those accomplished in empirical sciences. The involved variables are picked out; then through the sequence A one gets a device to reckon the time in the animation in a conventional way: namely the variable time is made explicit. Of course this subtle point is not so explicit for students: it is a practice induced by the instrumented actions of TI-Ns; students find natural to do so, since it works. It is interesting to observe that the scattered plot combines the time variable A Vs/ the length variable B or C. This practice is almost impossible or at least not simple in Cabri. The reason is in the possibility for TI-Ns of making the time variable explicit within mathematics itself.[6] That is, given a mathematical problem (like Problem 1) one can do an experiment very similar to those made in empirical sciences: one picks up the (supposed) variables that are important for the problem; makes a concrete experiment that involves such variables; then the mutual relationships among them are studied (using the scattered plot) and a mathematical model is conjectured and tested (e.g., through new experiments). In the end possibly the reasons why such a model is got are investigated, namely a proof of a mathematical sentence is produced. All this happen in a very precise method (picking up of variables, designing the experiment, collecting data, producing the mathematical model, proving), which is made palpable by different canonical functions of TI-Ns (respectively, giving a name to the variables, animation or dragging, data capture, either scattered plot and its adjustments, or regressions and scattered plot): only the last one, proving, is not supported in a direct way by a precise function of the software. In Cabri there are two main differences with respect to this situation: first, it is not possible to explicit the variable time through the "Newton trick"; second, it is not possible to collect data like in TI-Ns: as far as I know, the only way is through trace and geometrical locus, namely by-passing the numerical aspects. The two differences are crucial. In fact in Cabri it is possible to make experiments, but without "the time of the environment" (that is it is possible to have time as a variable that refers to some external event, for example for studying the graph of $y = t^2$, where t is time in an abstract sense, but that does not refer to the "time" of what is happening in the Cabri-world). Some consequences of the second difference, which consists in systematically bypassing the numerical aspects (that is the spreadsheet), will be discussed in §4.2. Here I underline that this bypassing destroys the perfect correspondence between the empirical method stressed above and some precise functions of TI-Ns. In this sense TI-Ns introduces new methods in mathematics. They do not only consist in the possibility of making explorations: this happens in many software, specifically in Cabri through the different types of dragging (Arzarello et al., 2002), in Derive through the use of moving cursors, etc.. In TI-Ns they consist in precise protocols that students learn to use and are very similar to the way external data concerning certain quantities are got through the use of probes connected to a computer. In our case the measures are collected through the "data capture" from the "internal experiment" made in the TI-Ns mathematical world (e.g., from the animation developed in G&G environment) to the Spreadsheet. From the one side, these methods are empirical, but from the other side they concern mathematical objects and computations or simulations with the computer and not physical quantities and experiments. Hence I propose the name of *almost-empirical* for them. The term recalls the vocabulary used by some scholars in the foundation of mathematics: for example Lakatos (1976) and Putnam (1975) claim that mathematics has a *quasi-empirical* status (for a survey of this issue see Tymoczko, 1986). I use the word almost-empirical to stress a different meaning: the main feature of the almost-empirical methods in TI-Ns is the list of the precise protocols that its users follow to make their experiments, like experimental scientists follow their own precise protocols in using machines for their experiments. In such a sense this method is somehow different from the more general description of quasi-empirical methods given by Lakatos or Putnam.

Almost-empirical actions made by students have also an epistemic nature, hence are not exclusively pragmatic; but they are specific of TI-Ns; hence I distinguish them from other epistemic actions that TI-Ns share with other software, e.g., exploring a situation with dragging or using a cursor for varying the parameters of an animation. An argument for this distinction is that in TI-Ns one can find all the epistemic actions produced within the other main didactical environments, but the converse is not true.

There are also strong similarities between instrumented actions produced in TI-Ns and Cabri environments: they are revealed analysing the students' actions, productions and communications according to the multimodal paradigm. Namely, when working with Cabri as well as with TI-Ns in small groups, students in both cases use a variety of semiotic resources and not only the ones specific of software: gestures, drawing, speech. In this sense TI-Ns results similar to other DGS. However, also in this case there is an important difference, whose cognitive and didactic meaning is discussed below. Namely, while in Cabri when students start writing their report, there is a strong break with the use of the instrument, in the sense that the exploration stops and a new phase starts where writing is prevailing in TI-Ns this cut is less evident: students generally write a first part of the report using the Notes sheet within TI-Ns, then come back to the other sheets of the instrument, make new actions and productions, and go back to the Notes. This probably happens because the Notes sheet is itself a part of the environment: the reporting phase is within the same environment as the experimenting one.

The presence of almost-empirical instrumented actions seems to have important consequences for learning. Typically, comparing the epistemic actions in DGS with the almost-empirical ones within TI-Ns, we find that in the first cases there is a big bridge between the perceptual level at which such actions (e.g., *wandering* and *dummy locus* dragging in Cabri: see Arzarello et al., 2002) are drawn and the theoretical level, which is necessary to develop proofs or to build up models of the studied phenomenon.

In the first case (exploring-conjecturing-proving theorems in geometry), the transition to the deductive level is generally marked by a cognitive shift from *ascending to descending* modalities,[7] according to which the figures on the screen are looked at: usually the shift is marked through an *abduction*, which also shows the transition from an inductive to a deductive approach. Generally speaking, the *tempos*, according to which such phenomena develop in time are long and require a high ingenuity by the pupils. The software is a support but the gap may be very large.

In the case of TI-Ns, there are two different evolutions, possibly because of the almost-empirical modalities of the instrumented actions drawn in this software. In the case of tempos, instead of a unique long episode with the transition from the ascending to the descending modality, generally there are many shorter episodes, each marked by the use of a different function of the software: first the data capture, then the scattered plots (which can be done without yet knowing which variable is function of which), then the

(a)

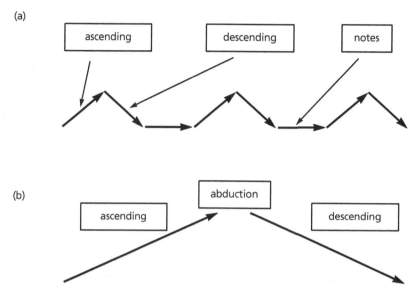

(b)

Figure 14.6 (a) TI-Ns tempos; (b) Cabri tempos

interpretation of the graph. Each action corresponds to very short alternation of ascending-descending modalities.

Moreover, because of the didactic contract, these shorter phases generally contain an epistemic part, where the students write in the Notes what they are doing, their conjectures etc, without breaking the almost-empirical process they are developing (a further big difference with Cabri). Instead of a unique long instrumented epistemic action, there are many short and different instrumented almost-empirical actions (see Figures 14.6a, 14.6b).

This breaking into smaller units makes easier for students to manage their practices in the new environment (once enough pragmatic actions have been developed in order to acquire a suitable knowledge of the instrument): the cognitive load is smaller. Moreover their smaller actions are instrumented through some precise specific functions of the software. This makes it possible to institutionalise more easily the corresponding practices and less ingenuity is needed.

It must be observed that TI-Ns requires more pragmatic actions to be learned at a level sufficient to use it as a suitable tool to solve problems without the support of the teacher. In the two classrooms of our teaching experiment 12 hours have been necessary. In this period the major part of student actions have been pragmatic, while only a minor percentage has been classified as almost-empirical or epistemic. After the apprenticeship period the students show a diminishing of the pragmatic actions and a sharp increasing of their epistemic and almost-empirical ones. We have also analysed the protocols of pupils that used Cabri in solving problems and that knew the software very well: the percentage of time that they spend in producing epistemic actions is comparable with the one that students using TI-Ns spend in producing almost-empirical + epistemic actions (with a proportion of 2 to 1 in favour of almost-empirical actions with respect to other epistemic ones). In both environments the epistemic and the almost-empirical actions have a strong perceptuo-motor nature: the main difference between the two consists in the different alternating shorter ascending-descending phases in TI-Ns with respect with longer one way ascending and descending phase in Cabri.

The impression is that such a difference stresses an attitude that tunes more with nowadays youth, who likes better a multitask way of operating with ICT, doing many things together within many modalities but for shorter periods of time than concentrating for a longer time within a unique modality (see the surfing way of life, described in Baricco, 2006). The almost-empirical actions with all their multimodal aspects and particularly their tempos feature some fresh practices that enter the classroom and are in positive resonance with practices within fresh representational infrastructures (for this notion see Kaput, Noss, & Hoyles, 2002) that are more and more diffuse in our society. The APCs built up by students using TI-Ns uses such modalities that come from the world outside the school (external practices). This may enhance the cognitive capabilities of our students in learning mathematics insofar it supports performances and modalities that do not require *ad hoc* learning.

4.2. Towards a Meaningful Use of Symbols

The second specific feature, which distinguishes TI-Ns from other software, is the instrumented actions that students develop using the symbolic spreadsheet of TI-Ns. This is a strong didactical innovation.

I will illustrate this issue through an example from the 9-th grade students, who were studying functions through their tables of differences. They had already learnt that for first degree functions the first differences are constant. They were asked to make conjectures on what functions have the first differences that change linearly.

Their conjecture was that quadratic functions have this property and arranged a spreadsheet like in Figure 14.7a, where they utilised:

- Columns A, B, C, D to indicate respectively the values of the variable x, of the function $f(x)$ (in B_i there is the value of $f(A_i)$) and of its related first and second differences (namely in C_i there is the value $f(A_{i+1}) - f(A_i)$ and D_j there is the value $C_{j+1} - C_j$);

(a)

x	f(x)	df(x)	ddf(x)	x0	a	b	c	h
0	3	-1	-4	0	-2	1	3	1
1	2	-5	-4					
2	-3	-9	-4					
3	-12	-13	-4					
4	-25	-17	-4					
5	-42	-21	-4					
6	-63	-25	-4					
7	-88	-29	-4					
8	-117	-33	-4					
9	-150	-37	-4					
10	-187	-41	-4					
11	-228	-45	-4					
12	-273	-49	-4					
13	-322	-53	-4					
14	-375	-57						
15	-432							

(b)

x	f(x)	df(x)	ddf(x)	x0	a	b	c	h
x0	a*x0^2+b*x0...	a*(h^2+2*h*x0)+...	2*a*h^2	x0	a	b	c	h
h+x0	a*(h+x0)^2+...	a*(3*h^2+2*h*x0...	2*a*h^2	x0	a	b	c	h
2*h+x0	a*(2*h+x0)^2...	a*(5*h^2+2*h*x0...	2*a*h^2					
3*h+x0	a*(3*h+x0)^2...	a*(7*h^2+2*h*x0...	2*a*h^2					
4*h+x0	a*(4*h+x0)^2...	a*(9*h^2+2*h*x0...	2*a*h^2					
5*h+x0	a*(5*h+x0)^2...	a*(11*h^2+2*h*x...	2*a*h^2					
6*h+x0	a*(6*h+x0)^2...	a*(13*h^2+2*h*x...	2*a*h^2					
7*h+x0	a*(7*h+x0)^2...	a*(15*h^2+2*h*x...	2*a*h^2					
8*h+x0	a*(8*h+x0)^2...	a*(17*h^2+2*h*x...	2*a*h^2					
9*h+x0	a*(9*h+x0)^2...	a*(19*h^2+2*h*x...	2*a*h^2					
10*h...	a*(10*h+x0)^...	a*(21*h^2+2*h*x...	2*a*h^2					
11*h...	a*(11*h+x0)^...	a*(23*h^2+2*h*x...	2*a*h^2					
12*h...	a*(12*h+x0)^...	a*(25*h^2+2*h*x...	2*a*h^2					
13*h...	a*(13*h+x0)^...	a*(27*h^2+2*h*x...	2*a*h^2					
14*h...	a*(14*h+x0)^...	a*(29*h^2+2*h*x...						
15*h...	a*(15*h+x0)^...							

Figure 14.7

- Variable numbers in cells E2, F2,..., I2 to indicate respectively: the values x_0 (the first value for the variable x to put in A2); a, b, c for the coefficients of the second degree function $ax^2 + bx + c$; the step h of which the variable in column A is incremented each time for passing to A_i to A_{i+1}.

Modifying the values of E2, F2,...,I2 the students can easily do their explorations. This is a learned practice, that gradually becomes a shared practice in the classroom, because of the interventions of the teacher, who stresses its value as an instrumented (epistemic) action, which supports explorations in the numerical environment.

It is interesting to observe that such a practice reveals its didactical power if analysed through a semiotic lens. Using the terminology in Duval (2006), this instrumented action supports a systematic *treatment* of numbers, scaffolded according to the formula of the second-degree function. This types of treatments, according to the frame of Duval, is one of the roots for developing algebraic thinking in students. Hence the instrumented actions of this type seem to be promising in the learning of algebra. In this TI-Ns is not different from usual spreadsheets and share with them such potentialities with respect to the learning of the algebraic language.

But, this is only the first half of the story. There is a more interesting second part, which happens because of the symbolic calculations that TI-Ns can support with its spreadsheet (a typical CAS feature).

In fact students realise that:

- If they change only the value of c, only column B changes, while the columns C and D of the first and second differences do not change; hence they argue that the fact that and the way how a function increases/decreases does not depend from the coefficient c;

- If they change the coefficient b, then columns B and C change but column D does not; many students conjecture that the coefficient b determines if a function increases or decreses but not its concavity;
- If they change the coefficient a, then columns B, C and D change; hence it is the coefficient a to be responsible of the concavity of the function.

A difficult point here is to understand why such relationships hold. The tables of numbers do not suggest anything. It is the symbolic power of the spreadsheet to be useful in this case. The epistemic instrumented action in this case is very interesting and consists in substituting letters to the numbers: see Figure 14.7b. This practice has been suggested by the teacher in most cases but a couple of students have done it autonomously. The spreadsheets show clearly in this case that the value of the second difference is $2ah^2$. In this case the letters condense the symbolic meaning of the explorations developed before in the numerical environment.

The teacher has stressed this power of the symbolic spreadsheet in the next lecture and again a fresh practice has entered the classroom.

In this sense TI-Ns allows an anticipated exposition of students to the symbolic aspects of the mathematical language supporting suitable instrumented actions, which are particularly apt to trigger the symbolic function of the algebraic language.

5. THE ROLE OF THE TEACHER

In all our discussion I have pictured mainly the relationships between the software and the students; the role of the teacher generally has been in the rear. But, our picture is only half of the story. Roughly speaking, without the teacher the software alone cannot mediate anything; hence our answers ad Q1, Q2 in a sense are wrong or at last incomplete. Our answers to the questions Q1–Q2 cannot be given in an absolute way: they are deeply determined by the role of the teacher in designing and managing the classroom situations with the software. Hence they are valid provided one embeds the activities with the software in a suitable didactical design and in suitable orchestrated interactions between the teacher and the students. In fact students' learning processes do depend not only on their interaction with the software but also and essentially on their interactions with the teacher. The space does not allow to elaborate this point properly and I will limit to sketchily discuss a major point.

The role of the teacher in promoting learning processes has been analysed according to different frameworks. According to a Vygotskyan conceptualization of ZPD (Vygotsky, 1978, p. 84), teaching consists in a process of enabling students' *potential* achievements. The teacher must provide the suitable pedagogical mediation for students' appropriation of *scientific concepts* (Schmittau, 2003). Within such an approach, some researchers (e.g., Bartolini & Mariotti, 2008) picture the teacher as a *semiotic mediator*, who promotes the evolution of *signs* in the classroom from the personal senses that the students give to them towards the scientific shared sense. In this case teaching is generally conceived as a system of actions that promote suitable processes of *internalisation*. I have pictured this approach within the APC-space model (see Figure 14.3).

Our semiotic lens allows framing and describing an important semiotic phenomenon, which lives in the APC-space, that I have called *semiotic game* (Arzarello & Paola, 2007) and that is particularly fruitful within technological environments (see the example discussed in Arzarello & Robutti, 2008).

Roughly speaking, the teacher develops a semiotic game with the students when (s)he coordinates with the semiotic resources used by the students and then guides the development of knowledge using these resources.[8] Typically when the student produces some non verbal sign (e.g., a gesture) that, according to the teacher judgment, shows that s(he) is in ZPD with respect to the knowledge to be taught, the teacher echoes the student using the same non verbal resource but uses a precise mathematical language to phrase it. Doing so, (s)he supports the students towards a correct scientific meaning. A typical example of semiotic game is when the student tries to interpret some sign on the screen, but while her/his non verbal resources show that possibly (s)he is understanding the meaning of that sign, her/his words are far from showing this. Then the teacher supports her/him borrowing her/him the right words that interpret her/his non verbal cues. The semiotic games practise is rooted in the craft knowledge of the teacher (Brown & McIntyre, 1993), and most of times is pursued unconsciously by her/him. Once explicit, it can be used to properly design the teacher's intervention strategies in the classroom for supporting students' internalisation processes.

Semiotic games are typical communication strategies among subjects, who share the same semiotic resources in a specific situation (namely they share their APC space). Through them the teacher can develop her/his *semiotic mediation*, which pushes students' knowledge towards the scientific one. Roughly speaking, semiotic games seem good for focussing further how "the signs act as an instrument of psychological activity in a manner analogous to the role of a tool in labour" (Vygotsky, 1978, p. 52) and how the teacher can promote their production and internalisation. The space does not allow to give more details and I limit myself to sketchily draw some didactical consequences. A first point is that students are exposed in classrooms to cultural and institutional signs that they do not control so much (e.g., the signs produced by a software). A second point is that learning consists in students' personal appropriation of the signs meaning, fostered by strong social interactions, under the coaching of the teacher. As a consequence, their non verbal productions, e.g., gestures, become a powerful mediating tool between signs and thought. From a functional point of view, gestures can act as "personal signs"; while the semiotic game of the teacher starts from them to support the transition to their scientific meaning. Semiotic games constitute an important step in the process of appropriation of the culturally shared meaning of signs, that is they are an important step in learning. They give the students the opportunity of entering in resonance with teacher's language and through it with the institutional knowledge. However, in order that such opportunities can be concretely accomplished, the teacher must be aware of the role that multimodality and semiotic games can play in communicating and in productive thinking. Awareness is necessary for reproducing the conditions that foster positive didactic experiences and for adapting the intervention techniques to the specific didactic activity.

This last chapter shows that the "scientific" answers to the naïf question put at the beginning of the paper make sense provided one is aware that they strongly depend on the didactical practices of the classrooms that is considered.

ACKNOWLEDGMENTS

Research program supported by MIUR and by the Università di Torino and the Università di Modena e Reggio Emilia (PRIN Contract n. 2005019721).

NOTES

1. In the research were involved: a colleague of mine, Ornella Robutti; a postdoc student, Cristina Sabena; a Master student, Silvia Damiano; and the two teachers, Domingo Paola and Pierangela Accomazzo. I thank here all of them. Of course the responsibility of what written here is exclusively of the author.

2. People working in cognitive science distinguish between pragmatic actions—actions performed to bring one physically closer to a goal—from epistemic actions—actions performed to uncover informatioan that is hidden or hard to compute mentally (see Kirsh & Maglio, 1994). In particular epistemic actions have been used by Dreyfus, Hershkowitz, and Schwarz (2001) in the context of mathematics learning. They define an epistemic action as a "mental action by means of which knowledge is used or constructed." Epistemic actions are often revealed in suitable settings, e.g., when students interact with an instrument.

3. We use the term *sign* or *semiotic resource* according to the Peirce definition, *namely as* anything that "stands to somebody for something in some respect or capacity" (Peirce, 1931–1958, vol. 2, paragraph 228). As such we can include as signs also resources not usually considered as belonging to semiotic systems in a strict sense.

4. See Kaput, Noss, & Hoyles (2002) for the notion of Representational Infrastructure.

5. Some questionnaires have been given to the students during the experimentation to get their evaluation and degree of approval of the software they were using.

6. This procedure is very similar to the way Newton introduced his idea of scientific time in science, distingushing it from the fuzzy idea of time, about which hundreds of philosophers had (and would have) speculated. For this reason I call this procedure the "Newton trick". Newton writes:

> "…eapropter ad tempus formaliter spectatum in sequentibus haud respiciam, sed e propositis quantitatibus quae sunt ejusdem generis aliquam aequabili fluxione augeri fingam cui caeterae tanquam tempori referantur, adeoque cui nomen temporis analogicè tribui mereatur. Si quando itaque vocabulum temporis in sequentibus occurrat…eo nomine non tempus formaliter spectatum subintellegi debet sed illa alia quantitas cujus aequabili incremento sive fluxione tempus exponitur et mensuratur." (Newton, 1969, p. 72)

> ("…for these reasons in what follows, I do not look at the time formally considered, but from proposed quantities, which are of the same genus, I suppose that one of them grows up with equable flux, to which all the others can be referred as if it were the time; so by analogy the name of time could be given to it with some reason. Hence, each time the word time will occur in what follows…by that word one will not mean the time formally considered, but that other quantity, through whose growth or equable flux the time is exposed and can be measured." Translation by the author).

7. The modality is ascending (from the environment to the subject) when the subject explores the situation, e.g., a graph on the screen, with an open mind and to see if the situation itself can show her/him something interesting; the situation is descending (from the subject to the environment) when the subject explores the situation with a conjecture in her/his mind. In the first case the instrumented actions have an explorative nature (to see if something happen), in the second case they have a checking nature (to see if the conjecture is corroborated or refuted). For more information, see Saada-Robert (1989), Arzarello (2000), Arzarello et al. (2002).

8. A more detailed description of semiotic games is in Arzarello & Robutti (2008); to be described properly a technical tool is necessary, namely the *semiotic bundle,* which is an enlargement of the notion of *semiotc system,* as defined in Ernest (2006).

REFERENCES

Artigue, M. (1997). Le Logiciel 'Derive' comme revelateur de phénomènes didactiques liés à l'utilisation d'environnements informatiques pour l'apprentissage. *Educational Studies in Mathematics 33*(2),133–169.

Arzarello F. (2000). *Inside and outside: Spaces, times and language in proof production.* Proceedings of PME XXIV, (vol. 1).

Arzarello F., (2008). Mathematical landscapes and their inhabitants: Perceptions, languages, theories. *10th International Congress on Mathematical Education, Plenary lecture* (Editor: M. Niss). IMFUFA, Roskilde University: Copenhagen, Denmark, 158–181.

Arzarello, F., Olivero, F., Paola, D., & Robutti, O. (2002). A cognitive analysis of dragging practises in Cabri environments, *Zentralblatt für Didaktik der Mathematik 34*(3), 66–72.

Arzarello, F., & Paola, D. (2007). Semiotic Games: the role of the teacher., *Proc. 31th Conf. of the Int. Group for the Psychology of Mathematics Education.* Seoul, Korea: PME.

Arzarello F., & Robutti, O. (2008). Framing the embodied mind approach within a multimodal paradigm. In L. English (Ed.), *Handbook of international research in mathematics education,* (pp. 720–749). New York: Routledge.

Baricco, A. (2006). *I barbari.* Fandango libri: Milano.

Bartolini, M. G., & Mariotti, M. A. (2008). Semiotic mediation in the mathematics classroom. In L. English (Ed.), *Handbook of international research in mathematics education,* (pp. 746–783). ISBN: 10:0-8058-5875-X. New York: Routledge

Brousseau, G. (1997). *Theory of Didactical Situations in Mathematics.* Dordrecht: Kluwer Academic.

Brown, S., & McIntyre, D. (1993). *Making sense of teaching.* Buckingham, UK: Open University Press.

Dreyfus, T., Hershkowitz, R., & Schwarz, B. (2001). Abstraction in context: the case of peer interaction. *Cognitive Science Quarterly 1,* 307–358

Duval, R. (2006). A cognitive analysis of problems of comprehension in a learning of mathematics, *Educational Studies in Mathematics* (61), 103–131.

Ernest, P. (2006). A semiotic perspective of mathematical activity: the case of number, *Educational Studies in Mathematics* (61), 67–101

Gallese, V., & Lakoff, G. (2005).The brain's concepts: the role of the sensory-motor system in conceptual knowledge. *Cognitive Neuropsychology 21,* 1–25.

Kaput, J., Noss, R., & Hoyles, C. (2002). Developing new notations for a learnable mathematics in the computational era. In L. D. English (Ed.), *Handbook of international research in mathematics education,* (pp. 51–75). Mahwah, NJ: Lawrence Erlbaum.

Kirsh, D., & Maglio, P. (1994). On distinguishing epistemic from pragmatic action. *Cognitive Science 18,* 513–549.

Lakatos, I. (1976). *Proofs and refutations.* UK: Cambridge University Press.

Maschietto, M., & Trouche, L. (to appear). Instrumental geneses and mathematics learning, social and historical aspects. *ZDM–The International Journal on Mathematics Education.*

Newton, I. (1969). *De Methodis Fluxionum et serierum*, In D. T. Whiteside (Ed.), *The mathematical papers of Isaac Newton* (Vol. III. p. 72). Cambridge, UK: At the University Press.

Peirce, C. S. (1931–1958). Collected papers, Vol. I-VIII. Cambridge, MA: Harvard University Press.

Putnam, H. (1975). What is mathematical truth? *Historia Mathematica* 2(1975), 529–543.

Saada-Robert, M. (1989). La microgénèse de la rapresentation d'un problème, *Psychologie Française*, 34(2/3), 193–206.

Schmittau, J. (2003). 'Cultural-historical theory and mathematics education'. In A. Kozulin, B. Gindis, V. S. Ageyev, & S. M. Miller (Eds.), *Vygotsky's educational theory in cultural context* (pp. 225–245). New York: Cambridge University Press.

Seitz, J. A. (2000). The bodily basis of thought, new ideas in psychology. *An International Journal of Innovative Theory in Psychology* 18(1), 23–40.

Tapscott, D., & Williams, A. D. (2007). *Wikinomics*. New York: Penguin Books.

Tymoczko, T. (1986). *New directions in the philosophy of mathematics*. Boston: Birkhauser.

Verillion, P., & Rabardel, P. (1995). Cognition and artifacts: A contribution to the study of thought in relation to instrumented activity. *European Journal of Psychology of Education* 10(1).

Vygotsky, L. S. (1978). *Mind in society*, Cambridge, MA: Harvard University Press.

Wilson, M. (2002). Six views of embodied cognition, *Psychonomic Bulletin & Review* 9(4), 625–636. [http://ecl.ucsd.edu/EmbCog_Wilson.pdf]